Center for the Study of National Reconnaissance Classics

THE HEXAGON STORY

CENTER FOR THE STUDY OF
NATIONAL RECONNAISSANCE
CHANTILLY, VA

APRIL 2012

Foreword

This volume re-publishes *The Hexagon Story* as part of the *Center for the Study of National Reconnaissance's (CSNR) Classics* series. The introductory information explains how this history of the Hexagon program focuses on the Air Force involvement with the program as it became operational and matured and contains limited discussion of the early Central Intelligence Agency (CIA) contributions to development of the program.

The history includes a 98-page annex that focuses on the role of the Intelligence Community in identifying collection requirements and the impressive results of that collection. The author of that annex, John Schadegg (a former Air Force officer and senior CIA official) was responsible for managing the collection operations and had first-hand insight. The annex has 57 pages of illustrative imagery examples, but all of the KH-9 panchromatic images are redacted because the primary film record remains classified at this writing in March 2012, pending review by the National Geospatial-Intelligence Agency. However, we included sixteen KH-9 panchromatic imagery products that the Director of National Intelligence declassified from this volume for the NRO to use during the turnover of Hexagon artifacts to the National Museum of the United States Air Force (NMUSAF).

The *Center for the Study of National Reconnaissance Classics* is a series of occasional CSNR publications whose purpose is to inform our readers about classic issues from the past. The books and monographs in the series most typically are histories, but they also could address lessons-learned topics, the legacy recognition of people and programs, insights into historically significant artifacts, or tutorials on the discipline of national reconnaissance. We issue the publications in the series on both an *ad hoc* basis, or in connection with a significant event. We are issuing a Gambit-Hexagon collection of histories in response to Director of the NRO Bruce Carlson's decision in June 2011 to declassify the programs and his subsequent declassification announcement on 17 September 2011. The Historical Documentation and Research (HDR) Section of the CSNR selected five classic histories of the Gambit and Hexagon programs:

- *A History of Satellite Reconnaissance—The Perry Gambit & Hexagon Histories* (by R. L. Perry)
- *The Gambit Story* (by F. C. E. Oder, J. C. Fitzpatrick, & P. E. Worthman)
- *The Hexagon Story* (F. C. E. Oder, J. Fitzpatrick, & P. E. Worthman)
- *Hexagon Mapping Camera Program and Evolution* (M. Burnett)
- *A History of the Hexagon Program—The Perkin-Elmer Involvement* (by R. J. Chester)

On 21 January 2012, the CSNR published the first volume in the Gambit-Hexagon CSNR Classics series, *A History of Satellite Reconnaissance—The Perry Gambit & Hexagon Histories*. We did this in support of the ceremony that marked the NRO turning over a collection of Gambit and Hexagon artifacts to the NMUSAF and their exhibit opening of these artifacts to the public. The opening of this exhibit represented the largest collection of satellite reconnaissance artifacts ever assembled and put on public display. That exhibit can serve as a companion resource to those who read the histories in this CSNR Classics collection.

Each of these histories offers a different perspective on the programs; the Perry Gambit and Hexagon histories are from the viewpoint of a former Air Force historian at RAND writing in response to tasking from the then NRO Program A (Air Force program); the Oder, et. al. Gambit and Hexagon histories are from the viewpoint of authors with program experience working under the sponsorship of the Deputy Director of the NRO; the Burnett Hexagon mapping system history is from the viewpoint of the Hexagon program office working under the direction of two Air Force officers in the program and the NRO Program A Director; and the Chester Hexagon history is from the viewpoint of Perkin-Elmer, which was an associate contractor for the Hexagon program.

All of the authors researched and wrote their histories during what some observers might describe as the height of the Cold War, from 1964 to 1985. This influenced them to react to and focus heavily on the threat from the former Soviet Union and its allies. Also, all of the authors had at least some degree of first-hand knowledge about these programs, and in many cases, they had first-hand experience working in the programs. This gives you a window into what it was like to be a participant-observer in the development and operation of these film-return satellite photoreconnaissance systems during the Cold War.

Dr. James D. Outzen, the NRO Senior Historian and Chief of the CSNR's HDR section, is the editor for the Gambit-Hexagon CSNR Classics series. Dr. Outzen selected the five histories for this CSNR Classics series from the NRO Records Center and CIA archives that collectively best retell the impressive Cold War story about these programs. He has prepared a brief preface and introduction for each history to provide context and explain its significance.

When you read the histories you will note that some information is missing. Even though the Director of the NRO authorized the declassification of almost all the programmatic information about these programs, some information, because of its potential impact on other sources and methods, remains classified. Dr. Outzen usually let the redacted text stand on its own, but in some instances he has done some editing for readability. For some of the histories, Dr. Outzen has incorporated supplemental reference material into the publication.

Robert A. McDonald, Ph.D.
Director
Center for the Study of National Reconnaissance

Preface

Coinciding with the commemoration of the 50th Anniversary of the National Reconnaissance Office (NRO), the Director of the NRO, Mr. Bruce A. Carlson, publicly announced the declassification of the Gambit and Hexagon imagery satellite systems on 17 September 2011. This announcement constituted the NRO's single largest declassification effort in its history. The Gambit and Hexagon programs were active for nearly half of the organization's history by the time of the declassification announcement. Their history very much represents the NRO's history—one that is defined by supremely talented individuals seeking state of the art space technology to address difficult intelligence challenges.

The United States developed the Gambit and Hexagon programs to improve the nation's means for peering over the iron curtain that separated western democracies from east European and Asian communist countries. The inability to gain insight into vast "denied areas" required exceptional systems to understand threats posed by US adversaries. Corona was the first imagery satellite system to help see into those areas. It could cover large areas and allow the United States and trusted allies to identify targets of concern. Gambit would join Corona in 1963 by providing significantly improved resolution for understanding details of those targets. Corona provided search capability and Gambit provided surveillance capability, or the ability to monitor the finer details of the targets.

For many technologies that prove to be successful, success breeds a demand for more success. Once consumers of intelligence—analysts and policymakers alike—were exposed to Corona and Gambit imagery, they demanded more and better imagery. Consequently, the Air Force, who operated the Gambit system under the auspices of the NRO, entertained proposals for an improved Gambit system shortly after initial Gambit operations commenced. They received a proposal from Gambit's optical system developer, Eastman Kodak, for three additional generations of the Gambit system. Ultimately the Air Force settled on only developing the proposed third generation because the proposed second generation offered minimal incremental improvement and the fourth generation appeared technologically unachievable at the time. The third generation became known as Gambit-3 or Gambit-cubed while it was under development. Once it replaced the first generation, it simply became Gambit. The new Gambit system, with its KH-8 camera system, provided the United States outstanding imagery resolution and capability for verifying strategic arms agreements with the Soviet Union.

Corona was expected to serve the nation for approximately two years before being replaced by more sophisticated systems under development in the Air Force's Samos program. It turned out that Corona served the nation for 12 years before being replaced by Hexagon. Hexagon began as a Central Intelligence Agency (CIA) program with the first concepts proposed in 1964. The CIA's primary goal was to develop an imagery system with Corona-like ability to image wide swaths of the earth, but with resolution equivalent to Gambit. Such a system would afford the United States even greater advantages monitoring the arms race that had developed with the nation's adversaries. The system that became Hexagon faced three major challenges. The first was development of the technology, which was eventually overcome by the Itek and Perkin-Elmer Corporations. The second was bureaucratic, deciding how the CIA and Air Force would cooperate in building such a system because they each had strengths and weaknesses in the development of national reconnaissance systems. The third challenge was to secure the resources that were required to build the most complicated and largest reconnaissance satellites at the time. By 1971, the NRO overcame the challenges to successfully launch the Hexagon satellite and fulfill, or even exceed, expectations for unparalleled insight into capabilities of US adversaries.

At the time of the Gambit and Hexagon declassification announcement, the NRO released a number of redacted Gambit and Hexagon documents and histories on its public website. One of the histories is contained in this volume.

The Hexagon Story was written in 1988 by Frederic Oder, James Fitzpatrick, and Paul Worthman. Since its publication in 1992, *The Hexagon Story* has served as a critical reference for the Hexagon program, alongside the work of Robert Perry. Oder, Fitzpatrick, and Worthman each had varied and rich backgrounds in Air Force national reconnaissance programs that provided a strong foundation

for researching and writing the histories of satellite imagery programs. They were asked by then NRO Deputy Director, Jimmie D. Hill, to write individual histories of the Corona, Gambit, and Hexagon systems. All three have since preserved the essential history of the programs.

The Hexagon Story is very rich in detail. The authors carefully document the Air Force's management of the Hexagon system once it was turned over to the Air Force program element at the NRO in 1973 by the CIA program element at the NRO. The authors include a wide range of summary tables and information including details of each launch, companies and personnel involved in the launches, photographs and illustrations, and the capabilities of the systems. The history is well-documented and sourced.

Since the authors' backgrounds are in national reconnaissance programs—and primarily in the Air Force element of the NRO—they offer unique insight into the Air Force's perspectives on the development, controversies, and management of the Hexagon program. If there is one shortcoming of this program history, it is the authors' minimization of the CIA's development and early management of the Hexagon program. Although both the Air Force and CIA elements at the NRO developed approaches for broad-area satellite search capabilities in the 1960s, the NRO eventually chose the CIA approach for what became the Hexagon program. Much of the CIA Hexagon story remains silent in this history.

The Hexagon Story joins five other volumes of Gambit and Hexagon histories that the Center for the Study of National Reconnaissance is reprinting in conjunction with the program declassifications. Those other volumes include *The Gambit Story* also written by Oder, Fitzpatrick, and Worthman, Robert Perry's histories of Gambit and Hexagon, a history of the Hexagon mapping camera, a Perkin-Elmer history of Hexagon, and a compendium of key Gambit and Hexagon program documents. In total, this collection of Gambit and Hexagon publications provides the public with broad insight into previously classified programs. The volumes complement each other in providing details not found exclusively in any single program history volume.

I have chosen not to reprint pages that were redacted in their entirety in *The Hexagon Story*. Those pages are: 149, 152, 156, 168 – 173, 180, 182 – 183, 187 – 204, 206 – 208, 210 – 211, 216, 223 – 224, and 226. We also did not reprint blank pages, which consist of pages vi, 6, 18, 108, 122, 126, 226, 236, 240, 244, 248, 254, and 260. The unedited redacted *Hexagon Story* can be found in the declassified records section of NRO.gov for those interested in reviewing a document with the completely redacted and blank pages.

The Gambit and Hexagon systems became reliable means for addressing difficult intelligence challenges once they became operational. The Hexagon systems, in particular, provided broad area imagery that was essential for understanding the strategic capabilities and arms control compliance of the Soviet Union and other Cold War adversaries. These national reconnaissance systems dutifully provided the nation reliable vigilance from above until the next generation of imagery satellites advanced the United States' intelligence collection capabilities.

James D. Outzen, Ph.D.

Chief, Historical Documentation and Research
The Center for the Study of National Reconnaissance

Center for the Study of National Reconnaissance

The Center for the Study of National Reconnaissance (CSNR) is an independent National Reconnaissance Office (NRO) research body reporting to the NRO Deputy Director, Business Plans and Operations. Its primary objective is to ensure that the NRO leadership has the analytic framework and historical context to make effective policy and programmatic decisions. The CSNR accomplishes its mission by promoting the study, dialogue, and understanding of the discipline, practice, and history of national reconnaissance. The CSNR studies the past, analyzes the present, and searches for lessons-learned.

~~Secret~~

~~NOFORN-ORCON~~
Handle via
BYEMAN-TALENT-KEYHOLE
Control Channels Jointly

The HEXAGON Story

~~Secret~~

BYE 140003-92
December 1992

Copy 12 of 26

WARNING: Information contained in this report is
classified, controlled, and eligible for foreign release
in accordance with the COMIREX Imagery Policy Series.

National Security Information

Unauthorized disclosure subject to criminal sanctions

Dissemination Control Abbreviations	NF (NOFORN)	Not releasable to foreign nationals
	NC (NOCONTRACT)	Not releasable to contractors or contractor/consultants
	PR (PROPIN)	Caution–proprietary information involved
	OC (ORCON)	Dissemination and extraction of information controlled by originator
	Rel	This information has been authorized for release to...
	WN (WNINTEL)	Warning notice–intelligence sources or methods involv

~~SECRET~~
~~NOFORN-ORCON~~

Contents

Section	Page
Preface	vii
1. Technological Ambush: A Nation at Risk	1
2. Technological Response: Scientists in the White House	3
3. Space Claimants and Inheritors	7
Space Claimant: The US Army	7
Space Claimant: The US Navy	8
Space Claimant: The US Air Force	8
The Primary Inheritor: A Solomonic Decision	11
Other Inheritors: The Department of Defense	12
4. The Air Force Space Heritage	13
Discoverer—CORONA	13
Accelerating the Samos Program	15
5. A New Inheritor: The National Reconnaissance Office	19
Formation of the National Reconnaissance Office	20
The Hazards of Organization	22
The Tyranny of Organizational Charts	23
An Open "Futures" Function	25
6. A New Space Claimant: FULCRUM	27
The Advent of Wheelon	27
The Advent of FULCRUM	29
Vance Sets Limits on FULCRUM	34
McCone Broadens the Limits on FULCRUM	35
Vance-McCone and System Engineering/Technical Direction (SE/TD)	37
7. Competing Claimants: FULCRUM and S-2	39
The Vance-McMillan Task Force and Steering Group	40
The Land Panel and FULCRUM	40
The Itek Episode	42
Perkin-Elmer Joins the FULCRUM First Team	44
The Land Panel Recommendations	45
Fierce Competition on an Uneven Playing Field	46
8. A New Space Inheritor: The CIA	53
A New DoD-CIA Agreement on the NRO	53
The Technical Task Group and the Project Management Task Group	56

~~SECRET~~
Handle via
BYEMAN-TALENT-KEYHOLE
Control Systems Jointly
BYE 140003-92

Contents (Continued)

The Report of the Project Management Task Group	57
ExCom Approves the HEXAGON Management Plan	64

9. The HEXAGON Development Program 65
- The HEXAGON Source Selection Gets Underway 67
- Early HEXAGON Development Activities 70
- Evolution of a Design Philosophy 75
- The Factory-to-Pad Process 76
- Development Progress 79
- System Description 81
- The Satellite Control Section (SCS) 81
- The Sensor Subsystem 83
- The Mapping-Camera Module 86
- The Donovan Review Committee 86
- Thermal Control 88
- Operational-Control Software 90
- Development Problems 90

10. The HEXAGON Flight Program 93

11. HEXAGON Under New Management 109
- Subsatellites and Experiments 116
- HEXAGON—A Unique Intelligence Asset 119

12. HEXAGON Financial Summary 121

13. A Goodly Heritage 123
- A Growing Technical Excellence 123
- A Growing International Acceptance 123
- A Founder's Accolade 125

Appendix A. HEXAGON and the Intelligence Community 127
- National Intelligence Requirements Management 127
- Flight Operations Management 129
- HEXAGON Imagery Security Policy 130
- System Requirements for the HEXAGON Photographic Reconnaissance System 131
- COMIREX Automated Management System (CAMS) 134
- National Imagery Exploitation Responsibilities 134
- Film Dissemination Responsibilities 136
- National Photographic Interpretation Center (NPIC) 136
- The National Imagery Interpretability Rating Scale (NIIRS) 137
- Weather Support—Key to HEXAGON Success 139

Contents (Continued)

HEXAGON Collection Requirements Formulation 144
Evolution of HEXAGON Broad-Area Search Requirements 144
Standing Search Delineations in the 1970s 148
Age of Search Imagery 153
HEXAGON Search Capabilities 153
HEXAGON Surveillance Capabilities 160
HEXAGON and Third World Countries 164
Color Imagery 164
HEXAGON Mapping, Charting, and Geodesy (MC&G)
  Capabilities 164
Epilogue 167
Imagery Illustrations 174
    Arms Limitations Agreements 174
    Detection 178
    Military Forces Order-of-Battle Information 181
    ██ 187
    Large-Scale Exercise Monitoring 192
    Nuclear Test Activities 195
    Economic Intelligence 197
    Natural Disasters 200
    Fortuitous Intelligence 203
    ████████████████████████ 206
    HEXAGON Quality 212
    ████████████████████████████ 224
  HEXAGON Success—A Team Effort 225
Appendix B. The CORONA Program 227
Appendix C. The GAMBIT Program 229
Appendix D. Agreement for Reorganization of the National
  Reconnaissance Program 231
Appendix E. HEXAGON and the Space Transportation System 237
Appendix F. Key Personnel on the HEXAGON Program 241

Glossary of Acronyms 245

Subject Index 249

Notes and References 255

~~SECRET~~
~~NOFORN-ORCON~~

Illustrations

James R. Killian, Jr.	4
George B. Kistiakowsky	16
Herbert "Pete" Scoville, Jr.	21
Albert D. "Bud" Wheelon	26
Edwin H. Land	28
John A. McCone	28
Brockway McMillan	28
Eugene G. Fubini	30
USA Lt. Gen. Marshall S. Carter	30
Cyrus Vance	33
Leslie C. Dirks	33
Perkin-Elmer FULCRUM Camera Optics	45
USN Vice Adm. William F. Raborn	47
Alexander H. Flax	52
Richard M. Helms	57
USN Capt. Frank B. Gorman	57
USAF Maj. Gen. John L. Martin, Jr.,	62
▬▬▬▬▬▬▬▬▬	66
USAF Col. Frank S. Buzard	66
Stanley I. Weiss	69
HEXAGON System Concept	71
HEXAGON Booster on Launching Pad	72
TITAN IIID Booster Vehicle	77
Factory-to-Pad Process	78
Hardware Flow	78
Satellite Vehicle Configuration	80
Satellite Basic Assembly Structure	80
Optical Bar Panoramic Camera Installation	84
Satellite Control Section	84
Mark 8 Reentry Vehicle	87
Mapping Camera Module	87
Film Transport	89
Film Path Schematic	89
Mapping Camera Module	91
SRV Inboard Profile	91
Mark V Reentry Vehicle	92
John L. McLucas	94
USAF Brig. Gen. Lew Allen, Jr.	96
William G. King, Jr.	96
Ascent, On Orbit, Deorbit Sequence	98
Aerial Recovery By C-130	98
Satellite Vehicle Assembly Going Vertical	99
Satellite Vehicle in Test Facility	100

~~SECRET~~
Handle via
BYEMAN-TALENT-KEYHOLE
Control Systems Jointly
BYE 140003-92

SECRET
NOFORN-ORCON

Illustrations (Continued)

USAF Col. Robert H. Krumpe	102
USAF Brig. Gen. David D. Bradburn	107
USAF Col. Raymond E. Anderson	107
USAF Brig. Gen. John E. Kulpa	112
USAF Col. Lester S. McChristian	112
Launching of Mission 1216, 17 June 1980	113
USAF Brig. Gen. Ralph H. Jacobson	115
USAF Col. Larry Cress	115

Reverse side blank

SECRET
Handle via
BYEMAN-TALENT-KEYHOLE
Control Systems Jointly
BYE 140003-92

HEXAGON Program History

Preface

This is the third volume in the history of the National Reconnaissance Program (NRP).

The first volume tells the story of CORONA—a program which was the initial application of space technology to the problem of carrying out overhead reconnaissance of denied areas. CORONA operated from 1960 to 1972. In its early days, it produced photographs with resolutions of 35-40 feet; however, the system was constantly improved and, by 1970, each CORONA mission was delivering several million square nautical miles (nm^2) of reconnaissance coverage at resolutions of 6-10 feet. CORONA served the nation well as a basic *search* system.

The second volume deals with GAMBIT, a system designed for the *surveillance* mode, necessarily covering less area than CORONA, but producing photography with a much better resolution. The system was operated from 1963 to 1984; it eventually achieved resolutions of ▓▓▓▓ or better, covering almost ▓▓▓▓ targets per flight.

By 1964, satellite reconnaissance technology had advanced to a point where it was predictable that *search* (CORONA) and *surveillance* (GAMBIT) modes might be combined within the capabilities of a single system. Studies of this possibility were undertaken under the auspices of the National Reconnaissance Office's (NRO) Program A (Air Force) and Program B (CIA), culminating in a decision to build a third major satellite system, called HEXAGON. This volume recounts the development and operation of HEXAGON, 1964-1986.

In preparing the manuscript, we appreciated the availability of several previously produced histories, as well as the presence of a number of key HEXAGON participants. In the former category, we drew on monographs by ▓▓▓▓▓▓▓▓▓▓▓▓▓▓▓▓ Col. Maurice G. Burnett, (USAF-Ret.),[‡] Donald E. Welzenbach,[**] and (once again) Robert Perry.[††]

Maj. Gen. John L. Martin, Jr., who headed the NRO Program A during HEXAGON's formative period was, as always, cordially helpful, as were Dr. Alexander H. Flax (Director, NRO, during HEXAGON's organizational phase), John N. McMahon (key member of the NRO Program B team), Stanley I. Weiss (first Lockheed Missiles and Space Company HEXAGON Program Director), and Walter Levison (a top official at Itek during the HEXAGON planning phase).

▓▓▓▓▓▓▓▓ "HEXAGON History," (draft), 29 Sep 73, BYE-107859-7.
▓▓▓▓▓▓▓▓ "Office of Special Projects, 1965-70," Vol. II, Central Intelligence Agency (internal) publication, Jan 73, BYE0-0400-72TS.
[‡] Maurice G. Burnett, Col. USAF-Ret., "HEXAGON (KH-9) Mapping Camera Program and Evolution," Dec 82, BIF-05W-23422.
[**] Donald Welzenbach, HEXAGON History (Working Paper), TKH-Byeman.
[††] Robert Perry, "A History of Satellite Reconnaissance," Vol. IIIB-HEXAGON, Nov. 73, BYE-17017-74.

Among direct contributors to the manuscript, we were fortunate to have the advice and personal assistance of Col. Frank S. Buzard (USAF-Ret.) and Lt. Col. John J. Schadegg (USAF-Ret.). Colonel Buzard, Program A's director of the HEXAGON Program from 1966 to 1971, joined us for working sessions in 1988 and prepared large segments of the early developmental and operational history. Colonel Schadegg, formerly chairman of the Imagery Collection Requirements Subcommittee of the COMIREX, was uniquely qualified to prepare Annex A, "HEXAGON and the Intelligence Community."

The basic manuscript profitted from close reading by, and the helpful comments of, Brig. Gen. Donald G. Hard, ██████████ and ██████████ (all of the NRO Staff), ██████████ (USAF-Ret., (formerly of the NRO Program A HEXAGON development office), Maj. Gen. John L. Martin, Jr. (USAF-Ret.), and Donald Welzenbach (until recently, a historian with the CIA). We are indebted to ██████████ and his associates at the National Photographic Interpretation Center (NPIC) for assistance in the selection of the HEXAGON photo-product; Dino A. Brugioni and ██████████ in particular, located outstanding examples of HEXAGON "take." Donald Welzenbach, once again, provided generous assistance in editing the final manuscript initially: ██████████ both of NPIC, directed the publication process.

Special mention must be made of the faithful cooperation of ██████████, NRO Staff, who located a special trove of key HEXAGON documents for us, and of Roger Marsh, who supported our needs at the CIA. ██████████ USAF, NRO Program A, furnished detailed financial information for the discussion at the end of the volume. We also wish to recognize the invaluable services of Betty Root whose faithful transcription of the authors' often illegible scrawls was truly an outstanding accomplishment.

Most fundamental of all, the need for this series of histories was envisioned by Jimmie D. Hill, Deputy Director of the NRO. We continue to have many occasions to be grateful for his sponsorship and guidance.

18 November 1988
Sunnyvale, California

Frederic C.E. Oder
██████████
Paul E. Worthman

Section 1

Technological Ambush: A Nation at Risk

On 20 January 1953, the international view from the East Front of the Capitol was far from reassuring. President Dwight D. Eisenhower, who had been sworn into office on that day, later reminisced:

> Two wars, with the United States deeply engaged in one, and vitally concerned in the other, were raging in Eastern Asia; Iran seemed to be almost ready to fall into Communist hands; the NATO Alliance has as yet found no positive way to mobilize into its defenses the latent strength of West Germany; Red China seemed increasingly bent on using force to advance its boundaries; Austria was still an occupied country, and Soviet intransigence was keeping it so. European economies were not yet recovered from the effects of World War II. Communism was striving to establish its first beachhead in the Americas by gaining control of Guatemala.[1]

The view was grim indeed; it had been darkened further by the shadow of a technological event: the testing of an atomic weapon by the USSR on 29 August 1949. This stunning achievement had occurred years ahead of the "probable schedule" predicted by US nuclear and intelligence experts; as a result, the international power structure was completely out of balance. The democratic nations had not prepared for such an emergency and could not make a coordinated response. And the next technological "ambush"—the Soviet production of a thermonuclear weapon—was already under preparation; the test of that weapon, in August 1952 (during Eisenhower's election campaign), occurred less than a year after the US counterpart.

There was more to come. At the time of Eisenhower's inauguration, the Strategic Air Command's inventory of "the bomber for tomorrow"—the B-52—consisted of two preproduction prototypes. Full deployment of the aircraft was planned for late 1955; but in May 1954, just 15 months after Eisenhower's inauguration, the deployment schedule was shadowed by the appearance of a Soviet intercontinental bomber called the BISON. The event was much more than a surprise; in actuality, it was an unprecedented threat, for the combination of a Soviet hydrogen weapon with an intercontinental carrier meant the United States was vulnerable to surprise nuclear attack.

For many years, the broad expanse of two oceans had provided a barrier to military assault upon the United States. During those years, the nation relished a thought that it had no "natural" or "dedicated" enemies. Over a period of a century and a half, major military actions in which the United States had engaged resulted either from internal dissension or from an ally's plea for assistance. Suddenly, invulnerability evaporated, and Eisenhower became the first President to carry the burden of this new concern.

On 27 August 1957, the Soviets announced the successful flight test of an intercontinental ballistic missile (ICBM). In achieving a third technological surprise, the USSR was a leader, rather than a follower. Although this was only a test, differing substantially from an operational capability, the effect of the Soviet action was dismaying. The US intermediate-range Thor missiles had attempted four highly publicized test flights (on 25 January, 19 April, 21 May, and 30 August 1957), with four failures; on 11 June, the first test flight of the Atlas ICBM had also failed. Immediately after the Soviet announcement of success, the expression "missile gap" came into American usage. The scope of national concern was reflected in Eisenhower's statement that "there was rarely a day when I failed to give earnest study to reports of our progress and to estimates of Soviet capabilities."[2]

On 4 October 1957, just five weeks after the ICBM bombshell, the Soviets placed Sputnik-I into orbit. One month later, before the world could catch its breath, Sputnik-II was launched, with a live dog as passenger and a television camera to prove it. On 6 December 1957, United States attempted to respond by launching the Navy's Vanguard satellite. Unfortunately, the vehicle malfunctioned and was destroyed by fire, on the launching pad, in full view of the American public, with television cameras grinding out the story. Americans had already heard the noun "crisis" associated with US bombers; then with US missiles; now the adjective would be "space." The series of technological surprises seemed endless and concern became general as the public wondered, "What next?"—half fearful of the reply.

Section 2

Technological Response: Scientists in the White House

Understandably, President Eisenhower's personal concern over "What next?" preceded public reaction. His own thoughts had been formulated during 1953—his first year in office—as he read National Security Council (NSC) studies, RAND Corporation reports, and Central Intelligence Agency (CIA) estimates which regularly specified each coming year as "the year of maximum danger,"[3] routinely vitiating their authority with the caveat: "Because of the absence of 'hard' intelligence data, our prognosis is the best that can be made, under adverse circumstances."

Like all national leaders, Eisenhower needed unequivocal answers to two questions: (1) What are our potential adversaries' capabilities? and (2) What are their intentions toward us? To answer the "intentions" query was concededly difficult, particularly in peacetime; however, the lack of a firm response to the "capabilities" question was intolerable.

In March 1954, Eisenhower met with Dr. Lee DuBridge, president of the California Institute of Technology, and the members of DuBridge's Office of Defense Mobilization Science Advisory Committee, to discuss these concerns and to solicit technological assistance in improving the quality and quantity of intelligence information on the USSR. DuBridge, in turn, asked Massachusetts Institute of Technology (MIT) President James R. Killian, Jr., to organize a subgroup to look into the matter. This group, in turn, suggested an in-depth examination of the nation's offensive and defensive potential. However, Killian realized that such a study required White House approval. Eisenhower, in July 1954, authorized the establishment of the Technological Capabilities Panel (TCP) to address the problem.

The TCP undertook its assignment energetically, addressing five formidable study areas, including, as the primary:

> Increasing our capacity to get more positive intelligence about the enemy's intentions and capabilities, and thus to obtain, before it is launched, adequate foreknowledge of a planned surprise attack.[4]

The subgroup working in this particular area was headed by Dr. Edwin H. Land, of the Polaroid Corporation. Just four months later (in early November 1954), Land's team recommended development of a very high-flying reconnaissance aircraft as the best immediate response to the "positive intelligence" problem. Impressed by the anticipated feasibility and capability of such a system, Eisenhower approved the development, "but he stipulated that it should be handled in an unconventional way so that it would not become entangled in the bureaucracy of the Defense Department or troubled by rivalries among the services."[5] Following his own guidance, the president assigned the U-2 project to the CIA, where Richard M. Bissell, Jr., Special

Assistant to the Director of Central Intelligence (DCI), set up a CIA-USAF development team.[6] The work proceeded swiftly: flight-testing began in August 1955, and the first overflight of the USSR occurred on 4 July 1956. The U-2 was used sparingly, discreetly, and successfully until 1 May 1960, when it was shot down by the Soviets while on a reconnaissance mission, thereby making its own unfortunate contribution to diplomatic and technological ambush. While the technological aspect was relatively minor—it had always been assumed that a Soviet fighter-missile threat would eventually challenge the overflights—the diplomatic consequences were disastrous, since the initially announced "cover story" was contradicted dramatically by the word and presence of a captured pilot.

James R. KILLIAN, Jr.

The USSR's premier satellite success, in 1957, made US leadership aware, rather abruptly, that it did not really have a space program; furthermore, that it had not even defined the objectives of such a program. Once again, there was an urgent need for scientific guidance at top governmental levels and, on 7 November 1957, Eisenhower appointed MIT's Killian to a newly formed position: "Special Assistant to the President for Science and Technology."

During his 20-month tenure, Killian met almost daily with the President. It was essential that he do so, for, with a nation in shock, and the national space scene unstructured and undisciplined, it would take continuing expert attention and effort to clarify top-level planning and to restore order. In addition, "space" assumed a very special importance, because it offered a possible solution to the basic dilemma studied by the TCP in 1954—55 ("Increasing our capacity to get more positive intelligence about the enemy's intentions and capabilities . . . ."); with the anticipated advent of ballistic missiles, the boosters essential to satellite reconnaissance operations would eventually be available.

"Clarifying top-level planning" meant deciding which space projects were truly essential to national welfare; "restoring order" required deciding which federal organizations should be assigned specific space tasks. On 7 February 1958, President Eisenhower approved a proposal made by Killian to centralize this effort. It was a new defense office—the Advanced Research Projects Agency (ARPA)—which would control, direct, and relate the military's missile and space programs. Secretary of Defense Neil H. McElroy implemented this organization over strong objection of the Joint Chiefs of Staff, who particularly disliked the word "direct." As a matter of fact, ARPA's scope was tremendous: for all practical purposes it was chartered to direct the national space program, since the National Aeronautics and Space Administration (NASA) did not yet exist (and NACA, as its title stated, was only a "National Advisory Committee on Aeronautics," and was not organized to produce hardware or manage large development programs). In spite of the services' protests, ARPA's mandate held firm; ARPA' first director, Roy W. Johnson (who had been a vice president at General Electric Company) essentially had McElroy's permission "to operate his agency as a 'fourth service'. . . within the Department of Defense.'" So, from February to October 1958, ARPA controlled the US space program, and became the initial "space inheritor" within the United States.[8]

"Space claimants" appeared immediately, each prepared to fight to the death for the right to rescue the nation from technological ambush and to assume an exclusive franchise for the crusade. Dr. Killian commented wryly that "given the complexity, hazards, and uncertainties of the space assignment, it is surprising that so many wished to take it on." [9]

Section 3

Space Claimants and Inheritors

Space Claimant: The US Army

In April 1946, the US Army—which at that time included the US Army Air Force—began a series of flight experiments at its White Sands Proving Ground in New Mexico, using captured German V-2 rockets. By October 1951, 66 of these rockets had been fired. In June 1950, the Army moved 130 German "Project Paperclip" rocket scientists from White Sands to Huntsville, Alabama, where, under the leadership of Wernher von Braun, work began on the design of battlefield missiles. In November 1955, Secretary of Defense Charles E. Wilson gave the Army responsibility for developing an intermediate-range ballistic missile (IRBM), the Jupiter, and, on 1 February 1956, the Huntsville organization was renamed the "Army Ballistic Missile Agency" (ABMA) and placed under the command of Maj. Gen. John B. Medaris.

The ABMA was soon locked in combat with Brig. Gen. Bernard A. Schriever's (Air Force) Western Development Division (WDD), to which the Secretary of Defense had also assigned development of an IRBM, called Thor. Later, in 1956, Defense Secretary Wilson announced that Thor had been selected as the US operational land-based IRBM; henceforth, the Army would be limited to developing missiles with ranges of 200 nautical miles (nm) or less. In spite of this severe jurisdictional setback, the ABMA immediately applied its impressive in-house talent to "hurriedly convert their Jupiter-C reentry test vehicle, an elongated Redstone topped by clustered solid-propellant upper stages . . . into a satellite launcher;"[10] on 31 January 1958, this vehicle became America's first successful entry in the space race. On the basis of this accomplishment, the Army began to lobby strenuously for a more comprehensive franchise. Killian, who had to listen to Army presentations frequently, observed:

> Having launched our first satellite, the Army's was an aggressive contender for the job. Medaris and von Braun campaigned with fierce religious zeal to obtain a central role in space for the Army. Medaris vehemently proclaimed that military satellites should have greater priority than ballistic missiles, that the space program rightfully belonged to the Department of Defense, and that it would be a terrible mistake to give responsibility for the US space program to an independent civilian space agency. He did not attack the establishment of ARPA, as did the Air Force, because he saw a chance that ARPA in its partnership with the Army could get and manage the space program.
>
> As I look back on his fight for the Army's space team, I can't help but be impressed by General Medaris's artful campaign, even though I could not approve of his methods and sought to thwart them.[11]

Later, it would be recalled that even in these very early days of the space era von Braun was speaking earnestly of a "dream booster"—a clustered-engine vehicle designed to deliver one million pounds of thrust.

In October 1957, the Army proposed a military reconnaissance satellite to the Department of Defense DoD; it was to use television cameras and "cover" the USSR every three days. Then in November, the Army pressed its case for a satellite defense system, advising that a program for developing such a weapon had been under intensive study at ABMA for some time.

Space Claimant: The US Navy

The US Navy, and particularly its Naval Research Laboratory (NRL) at Anacostia, just south of the Nation's Capitol, had shown strong leadership in space science, joining enthusiastically in the White Sands V-2 program. It had also pioneered in the use of balloon-launched sounding rockets, which typically involved a polyethylene balloon to lift the device to about 70,000 feet, where the rocket would ignite and boost an instrumentation package about 40 miles into space. When the supply of V-2s dwindled, the Office of Naval Research (ONR) sponsored the design of a new sounding rocket—the Aerobee—at John Hopkins University; this booster was followed by the larger Viking, which could reach an altitude of 136 miles. Unlike the Army, the Navy did not attempt to assemble an in-house capability for rocket manufacture.

In 1955, the Navy began preparing Project Vanguard, which was to be its contribution to the 1957 International Geophysical Year. Vanguard would use a Viking first stage and an Aerobee-Hi (improved Aerobee) second stage to place geophysical instruments into earth orbit. Although it was a modest project, in the autumn of 1957 it suddenly came into the limelight as a desperate response to the success of Sputnik. Unfortunately, during its widely advertised launching on 6 December 1957, the vehicle's first stage exploded, and the rocket collapsed on the pad. (It is noteworthy that Project Vanguard had been expressly forbidden the use of a *military* booster.) This experience had a strong adverse effect on Navy enthusiasm for making a "space claim;" however, during 1958, the Navy Bureau of Aeronautics did propose a manned space-flight vehicle. Called MER-1 (Manned Earth Reconnaissance), the plan featured a reentry vehicle that could be controlled from booster burnout to water landing.

Space Claimant: The US Air Force

In 1948, less than a year after the Air Force was established, Vice Chief of Staff Hoyt S. Vandenberg announced official Air Force doctrine: "USAF, as the service dealing primarily with air weapons—especially strategic—has logical responsibility for the satellite."[12] The satellite he referenced was, of course, a military satellite. Late in the year, the DoD's Research and Development Board reinforced the Vandenberg dictum by designating the Air Force as the single service authorized to fund studies of satellite vehicles.

In May 1946, well before these pronouncements, RAND—then a division of Douglas Aircraft Corporation—had published the results of its first study of "the satellite:" "Preliminary Design for an Experimental World-Circling Space Ship." This extensive report evoked Air Force interest, but, absent the powerful boosters which the "ship" would require, the interest was essentially academic. In November 1950, RAND recommended that the Air Force begin research on *reconnaissance* satellites to evaluate their feasibility and military utility; in addition, it volunteered to conduct such a study, if requested to do so. Because of a heightening US awareness of the strengthening USSR military potential, RAND's offer was accepted.

In 1953, the newly established Air Research and Development Command (ARDC) sponsored a follow-on RAND satellite study, titled "Project FEEDBACK." The study involved hundreds of participants in an exhaustive review of then-current speculation on satellite reconnaissance. In spite of massive technical detail, FEEDBACK findings could be summarized nicely into three basic postulates: (1) begin *now*, (2) it will cost $165 million, (3) it will take seven years. This final report was delivered to ARDC Headquarters on 1 March 1954. In May, concurrent with Eisenhower's decision to build ICBMs, ARDC was directed to study the possibility of translating FEEDBACK into reconnaissance hardware. The task was assumed by ARDC's Wright Air Development Center (WADC), which set up a small study team, supported by engineering groups at RCA, Martin, and Lockheed. This activity was called WS-117L, the "Advanced Reconnaissance System."

It was obvious that the actual development of a reconnaissance spacecraft could not outpace the development of its booster; it was also clear that WS-117L would require Atlas-class boosters. As a hedge against possible WS-117L pressures on, and incursions into, his Atlas development-production program, General Schriever, commanding the WDD in Los Angeles, recommended in 1956 that FEEDBACK applications be transferred to his organization. Schriever's action was essentially defensive: with Atlas, Titan, and Thor developments under way, his basic desire was to protect, rather than expand, his franchise. But the cadre that arrived at WDD to handle the project consisted of bright, hard-driving enthusiasts; by April 1956 they had an approved development plan in hand, and, by October, WDD had awarded a contract to Lockheed for WS-117L, which was renamed Project Pied Piper. The funds available to the program totaled $3 million.

Elsewhere in the ARDC, the prospect of new opportunities in space technology and satellite systems was a heady stimulant. Each of the ARDC's many centers was convinced that it could show cause, or a unique need, to become the "Air Force Space Center." If the ARDC could just seize the "space initiative" for the entire DoD, there would be new projects aplenty for all centers. It was pleasant to dream further: perhaps the *US* Space Center could be at Rome, or Holloman, or Albuquerque—each of which was suffering from a paucity of "important" projects. And although these competitive ambitions were divisive, the centers were united in one thought: new mission or no, there must never again be another WDD! That organization, with its high priority, ready cash, and direct command lines to the Secretary of the Air Force, should forever remain anathema to the "regular" ARDC.

The word "space" swept through ARDC like a virus; every project officer became an enthusiast, anxious to prove that the work sponsored by his office—if examined closely—was "*space*-oriented." At ARDC headquarters, the assistant commander for technology announced that his long list of ongoing projects, many of them older than the command itself, was already 62 percent "space-oriented;" it was a little embarrassing when "closer examination" prompted his staff hurriedly to move "rocket engine technology" from 27th to first place on its own "Propulsion Project Priority List."

And then there were the "space warriors," with their vision of space as a battlefield; they presented elaborate plans for defending the cislunar arena from unfriendly terrestrial forces. Dr. Killian summarized his reaction to these folk:

> The Air Force fought just as hard as the Army for the space assignment. The atmosphere and outerspace were a continuum, it [the Air Force] maintained with considerable logic, and it already was well advanced with an intercontinental ballistic missile program . . . .
>
> This was an impressive case, but it might have been stronger if the Air Force had suppressed some of its own special brand of fantasies about space. Its top-ranking officers freely predicted that the next war would unquestionably be fought with space weapons, and some of the smaller Air Force fry had visions of space wars and dropping bombs from satellites.
>
> It was strange now to recall the fantasies that Sputnik inspired in the minds of many able military officers. It cast a spell that caused otherwise rational commanders to become romantic about space. No sir, they were not going to fight the next war with weapons of the last war; the world was going to be controlled from the high ground of space.[13]

It was instructive to contrast the self-serving, franchise-oriented presentations described by Killian with a proposal prepared by an objective "outsider"—for instance, an organization which already had more than enough "orders" on hand and was capable of working dispassionately on the space "problem." General Schriever's busy WDD was such a unit; ARDC Headquarters asked WDD to prepare a Space System Plan which could serve as the system portion of a much broader Space System and Technology Plan, already under preparation at the command headquarters. Schriever responded at once with a terse, lucid proposal covering three realistic purposes for military space systems: reconnaissance, communication, and manned space flight. His proposal evaluated these tasks as feasible, the costs for start-up ($26 million) as reasonable, and the goals as explicit. ARDC Headquarters staff received the proposal, scanned it, and quietly locked it away, sending to the Pentagon, instead, its own vast "Astronautics Package."

William M. Holaday, the "missile czar" of the DoD, received the "package" on 24 January 1958. He reviewed the document—a five-year plan covering such exotica as "Manned Space Station" and "Manned Moon-Base"—which called for the *early* release of $1.7 billion. Following ARDC's example with WDD's plan, Holaday quietly locked away the "package." By 28 February 1958, even the Air Force Weapons Board had dropped the "package" from its future funding list.

The Primary Inheritor: A Solomonic Decision

In order of possible precedence, based on program strength and experience, the nation's claimants to space technology and operations were the Air Force, the Army, and the Navy. An additional claimant, in fourth place, could have been the NACA, but NACA had elected to abstain from the race.

> By early February 1958, as the Eisenhower administration began wrestling with the complexities of formulating a national space program for space exploration, NACA had taken the official position that with regard to space it neither wanted nor expected more than its historic niche in Government-financed science and engineering . . . . This would involve a continuation of NACA's traditional function as a planner, innovator, tester, and data gatherer for the Defense Department and the missile and aircraft industry.[14]

But there were strong external pressures for changing the character (and charter) of NACA. In October 1957, the American Rocket Society had called for a civilian space (research and development) agency. In November, the National Academy of Sciences endorsed a "National Space Establishment" to be organized under civilian leadership. In January 1958, Lyndon B. Johnson's Senate Preparedness Committee recommended establishing a national space agency and, by April 1958, there were 29 bills and resolutions in Congress relating to a national space effort.

Clearly, the time had come for a decision on organizing US space work, and only the President could moderate such an issue. The Eisenhower response was grounded on a fundamental conviction he had held since 1954 (when the nation had planned its contribution to the International Geophysical Year): *space activities should be peaceful activities.*

Early in 1958, Eisenhower asked Dr. Killian to make recommendations on an organizational model for the US space effort. Killian, who also chaired the President's Science Advisory Committee (PSAC), responded swiftly and categorically: NACA should be restructured and rechartered to become the focus of astronautics for the United States; such an arrangement would demonstrate, beyond doubt, the peaceful purposes and intentions of the nation. In April, Eisenhower forwarded Killian's recommendation to Congress, and on 29 July 1958 he signed the National Aeronautics and Space Act into law.

Thus NACA, which had never pressed its case as a Space Claimant, became National Aeronautics and Space Administration (NASA)—the nation's primary Space Inheritor. Along with this changeover came a substantive legacy: NASA was given the Navy's Vanguard, the Army's ABMA and Jet Propulsion Laboratory (at the California Institute of Technology), and a number of Air Force advanced technology programs (including the 1.5-million-pound thrust F-1 rocket engine subsequently used on the first-stage booster of the Apollo Moon Program) together with $117 million from DoD funds. More importantly, NASA acquired the national charter for manned space flight technology and operations. NASA—which had claimed the least—was granted the most: space science, space exploration, manned space flight, and planetary exploration.

Other Inheritors: The Department of Defense

NASA's acquisition of ABMA muted the Army's most vocal space claimants. As time went on, residual Army "space requirements" could—and would—be satisfied by access to DoD communication, geodetic, and reconnaissance satellites. Similarly, loss of the Vanguard team focused Navy space needs on communication, navigation, and reconnaissance satellites.

As for the Air Force, the President's largesse toward NACA was a stunning blow. There had been a constant (and reasonable) assumption, on the part of the Air Force, that any man in space would be blue-suited and that NACA would have, at most, a responsibility for advisory technical assistance to the Air Force. But, henceforth, the relationship would be exactly reversed: the Air Force would *assist* NASA, with launching services, tracking services, injection into orbit, and sometimes would even furnish the astronaut. But the programs themselves would belong to NASA.

There was, of course, one major assignment remaining. Toward the end of 1958, ARPA, which had controlled all military space programs since February, surrendered the "Advanced Reconnaissance System" to the Air Force. Similarly, ARPA transferred control of Transit (a navigation satellite) to the Navy and Courier (a communication satellite) to the Army.

Section 4

The Air Force Space Heritage

The year 1958 would always be commemorative for both the (new) NASA and the (somewhat new) Air Force. The division of the national space legacy had been made between "civil space" and "military space." The former, involving space science, space exploration, space stations, and planetary probes, had been awarded to NASA. The DoD would concentrate henceforth on the military uses of space: specifically, on space as an observation post, a communication center, and an arena for deterrence. Communication satellites would be typified by the Navy Transit and the Army Courier developments; the Air Force's Samos[13] (formerly called WS-117L and Sentry) would attempt to establish an observation post and its (Saint) inspector satellite would become a first step toward creating a space-based deterrent capability.

Discoverer-CORONA

A second milestone in enhancing military space technology occurred on 22 January 1958, when the NSC issued Action Memorandum No. 1846, which directed the DoD to give priority to the development of an operational reconnaissance satellite. The directive was very good news, particularly to the small group of officers still working on Sentry (later called Samos) at the Air Force Ballistic Missile Division (AFBMD formerly the WDD). By February 1958, Presidential Science Adviser Killian was convinced that the most promising *immediate* response to the NSC memorandum would be a "quick-fix" within the existing Samos program. Samos had been on "low-burner" at AFBMD, awaiting the availability of the mandatory ICBM-class booster—probably Atlas. Killian speculated that a reasonable Samos "quick-fix" could consist of a simpler, lighter payload than the existing Samos design—something that could be lifted into orbit by the already-available Thor IRBM.

There were other encouraging elements in such a proposal. A spacecraft (later called Agena) was sufficiently developed to be available to this "quick-fix system;" reentry vehicles could be crafted rather rapidly, using ICBM-originated technology; a global satellite-control network would soon be in existence to support in-flight operations; a spaceworthy camera was available; and an existing capability for aerial recovery of film payloads could be used (in lieu of the more sophisticated—but not yet developed—readout hardware of the original Samos scheme). Finally, security considerations could be satisfied by calling the "quick-fix" system Discoverer and advertising it as an exploratory precursor to Samos and Midas—a system needed to provide basic technical design data for reconnaissance successors. Publicly, Discoverer would continue to look like part of the Air Force space legacy; in private, it would have a "black" name—CORONA—and would move out of the Air Force and near the Office of the President of the United States—certainly an *ultimate* Inheritor!

Eisenhower agreed immediately to the Discoverer proposal; the need for reconnaissance information was so urgent that the idea was worth a gamble. With regard to a leader for this work, Killian and the President both thought of Richard Bissell, who had co-managed (with Air Force Col. Osmund J. Ritland) the U-2 project and had won Killian's accolade as "a brilliant project engineer."[16] On 7 February 1958, the assignment was made. It seemed reasonable, on all counts, to recall Bissell and Ritland to "special duty" at the summit. Ritland, now a brigadier general, was vice commander of the AFBMD, where Discoverer was already under development. He could readily and easily direct an enhanced priority and support level for Discoverer contractors and Air Force units. Bissell could handle any "black" contracting (essential to the camera development at Itek[17]); he could also provide a proper security system to protect the CORONA mission. The Discoverer-CORONA development officer would be Lt. Col. Lee Battle, who was in charge of Discoverer at the AFBMD. Battle's mandate would be extended to make him "agent for all interested components of the Government."[18] Bissell would strongly influence system progress at the same kind of monthly suppliers' meetings he and Ritland had used successfully in the U-2 development; further, Bissell would again be the basic governmental contact with Killian and the President himself.

With Bissell resuming his function as a "White House Project Officer," it might have been presumed that the CIA had emerged as the latest Space Inheritor. But this was not the case; Discoverer-CORONA continued, at least for the time being, under the aegis of ARPA. Discoverer had been assigned, previously and categorically, to the Air Force—by ARPA. CORONA was something new, but still under ARPA control. Rather than assigning CORONA per se to a military department, it was assigned to two persons—Bissell and Ritland,[19] who assumed their roles as *individuals*, fortuitously having advantageous authority within their more obvious jurisdictions. General Schriever, Ritland's "normal" supervisor, understood and supported the arrangement completely; Director of Central Intelligence (DCI) Allen Dulles, Bissell's supervisor, was, at this stage, in late career and did not pay much attention to "details of what was going on in his agency;"[20] he expected Bissell to proceed sagaciously and upon his own initiative.

At this same time, the Air Force was directed, by the Secretary of Defense, to streamline the administration of its satellite developments. In March 1958, the Vice Chief of Staff issued a memorandum, "Space Projects Involving ICBM/IRBM Components," which stated that channels and procedures identical to those of the ballistic missile program ("Gillette Procedures") would now be applied to space systems. For the AFBMD space system office, this meant that communication with USAF Headquarters could legitimately bypass the parent command (ARDC) and the Air Staff, going directly to the Office of the Air Force Chief of Staff. Six months later, ARDC Headquarters announced, somewhat redundantly, that it would assign any new space missions it might receive to AFBMD. Ironically, ARDC was already "losing" space systems (in the sense of "ownership") rather than "receiving" them.

At the time of Discoverer-CORONA's birth, the entire space system group at AFBMD was small: the professional and clerical staff numbered 52 and the officer-in-charge was a colonel. There were valid reasons, of course, for limiting the size of the office; Samos planning was restricted by the unavailability of Atlas boosters (defense priorities still dictated that all early ICBM production should go directly to Strategic Air Command operational sites). There was an additional restriction: the readout system envisioned for Samos had to be superior to existing state of the art. Furthermore, Samos was depending on the availability of space environmental information from early Discoverer flights as an aid to designing proper sensors and control equipment. Finally, the space office, as a relative newcomer to AFBMD, stood in the shadow of the ballistic missile monolith; strategic missile urgencies quite naturally diminished the priority of the newly arrived space systems.

Accelerating the Samos Program

It was not until 1960 that two events combined to shift priorities in favor of the Air Force space program. The first was the shootdown of a U-2 by the Soviets on 1 May 1960. With the cancellation of further reconnaissance flights, the United States lost its most precious source of (limited but vital) information on military installations and hardware in the USSR. The second event was the success of the CORONA program's Discoverer-XIV on 19 August 1960 (with "success" measured in terms of delivered exposed film).[21] The flight answered some crucial questions that had plagued Samos engineers: No, there were no serious equipment-disabling radiation effects; no, the electronic assemblies did not become erratic; no, the photographic film did not curl and crumble; yes, the pictures were excellent; yes, space was a feasible reconnaissance environment.

Eisenhower reacted immediately to CORONA's success. Shortly after the U-2 shootdown, he directed his new Science Adviser, George B. Kistiakowsky, to set up a study group to recommend alternative options to reconnaissance aircraft overflight. He now repeated his direction and, on 25 August 1960, six days after the CORONA success, Kistiakowsky responded. He recommended that Samos be given a streamlined management structure within the DoD—one possibly modeled on the CORONA program: ". . . the organization should have a clear line of authority and . . . on top level the direction [should] be of a national character, including the Office of the Secretary of Defense (OSD) and CIA . . . ."[22] Kistiakowsky observed that the comparable office for locating a Samos "management summit" would probably be the Office of the Secretary of the Air Force. This designation would place Samos management out of reach of both the ARDC and the Air Staff. In addition, management procedures would be as simple as possible, perhaps even more streamlined than those devised for the ballistic missile program.

On 1 September, the NSC directed the Secretary of Defense to set up such a Samos organization, consisting of two parts: the Secretary of the Air Force would have, on his personal staff, an office called SAFMS ("Secretary of the Air Force/Missiles and Space"); in the field, at Los Angeles, he would have SAFSP ("Secretary of the Air Force/Special Projects"), to manage the actual development of Samos. Thus the Secretary of the Air Force's office became, in literal fact, a research and development organization.

Henceforth, there would exist two Air Force space legatees: AFBMD and SAFSP, collocated in Los Angeles. AFBMD would retain remnants of the original Air Force inheritance: it still had Midas (an attack-alarm system), Vela-Hotel (a nuclear-detection satellite), and Saint (a simple satellite inspector); and it hoped to be assigned a communication satellite. But looking over and above this limited list, AFBMD could not help acknowledging that space systems with the most prestige, the greatest growth potential, and the largest cash flow had moved across the street, to the new SAFSP.

George B.
KISTIAKOWSKY

There was, of course, work for AFBMD to do in serving other agencies. In 1959, the Air Force had been made responsible for furnishing "booster-support services" to the Army, Navy, and NASA. These services covered a wide and expensive range of activity that *might* include the booster first stage (usually a Thor or Atlas), the second stage (an Agena or Able-Star), the final stage vehicle, total system engineering, procurement services for the system, a launching pad, launching services, injection into orbit, on-orbit command and control, and capsule recovery. Although the limited space assignments of the Army and Navy constrained their booster requirements, NASA, in its earliest years, had a continuous, extensive need for such support. In January 1961, the Wiesner Report, which examined the national space effort for newly elected President John F. Kennedy, observed that "the USAF provides 90 percent or more of the resources and physical support required by the space programs of other agencies." But supporting other agencies, while vital and worthwhile, was not the same as having one's own space projects. And, over the long haul, NASA, the big booster customer, would surely develop its own resources; its call for neighborly assistance was ephemeral. Only CORONA and Samos—both outside the AFBMD domain—could be depended upon as steady booster customers.

There was a sardonic coda to the "Inheritor scene" in March 1961, when DoD Directive 5160.32 appeared, stating that "research, development, test and engineering of Department of Defense space development programs or projects, which are approved hereafter, will be the responsibility of the Department of the Air Force." Later, reminiscing on this event, Secretary Eugene Zuckert observed that "it was like getting a franchise to run a busline across the Sahara Desert."[25]

As for the reconnaissance-satellite program, it had made a restless journey within the Air Force. The original Samos studies had been sponsored by Air Force Headquarters (1946-54); passed to ARDC for analysis (1954); sent to the WADC for detailed study (1954); transferred to the WDD for development (1956); with part of the task "lost" to ARPA and the Office of the President (1958); and the remainder going to the Office of the Secretary of the Air Force (1960).

ARDC, had never really "owned" CORONA; now it no longer owned Samos. Its new commander, Lt. Gen. B. A. Schriever, was one of the few persons in that headquarters to have a comprehensive knowledge of the forces and events that had reduced the AFBMD space mission to proprietary fragments and a multitude of "support" functions. Schriever's reaction, perhaps born equally of frustration and hope, was to separate the space residue from AFBMD and to request creation of a new organization: the Space Systems Division (SSD) ("Systems" could be pluralized because there were three of them). Perhaps a major general, as commander of SSD, would symbolize ARDC hopes and intentions; perhaps increased "exposure" of the residual space activity would attract the notice of DoD officials and help to reverse some recent high-level decisions. So, in April 1961, in the midst of mission program decline, a new division was born and encouraged to become more noticeable, more extensive, and more expensive.

Across a Los Angeles street from AFBMD, Brig. Gen. Robert E. Greer, newly appointed head of SAFSP, had a radically different view of mission and methodology. From the start, he was firm in his intention to keep his development organization as small, obscure, and cost-conscious as possible. He believed his mission was to examine, re-orient, and construct a reconnaissance system quietly, quickly, and reasonably.

Reverse side blank

Section 5

A New Inheritor: The National Reconnaissance Office

Presidential Science Adviser Kistiakowsky's delight in "protecting" satellite reconnaissance developments from the Air Staff and the ARDC was reflected in his journal entry for 25 August 1960: "If the Defense Department really sticks by its agreement with our recommendations on Samos, which will now be reinforced by an NSC directive, this may be the major accomplishment of my eighteen months in office."[24]

Samos' protective shield was soon extended further. James H. Douglas, Jr., Secretary of the Air Force, delegated his Samos responsibilities to Under Secretary Dr. Joseph V. Charyk (formerly Assistant Secretary of the Air Force for Research and Development). The newly organized SAFMS, directed by Brig. Gen. Richard Curtin, would be Charyk's personal Missile and Space staff; SAFSP, in Los Angeles, would be Charyk's field organization. There would be a minimum of formal communication between Charyk/Curtin and Greer; letters and memoranda would be replaced by cryptoteletype and KY-9 telephone.

On the West Coast, Greer had assembled a small, carefully selected cadre of officers who would assist him in reviewing the elements of Samos and devising ways to accelerate development progress. The term "Samos" had originally embraced six reconnaissance capabilities; Samos was a *family* of satellites, each of which was to be more sophisticated than CORONA. Samos would culminate in a version using read-out technology, rather than film recovery, for delivering reconnaissance photography. Developing all the "forms" of Samos was well understood to be a formidable task.

Given the pressure for a sophisticated reconnaissance system, Greer saw no gain in proposing the jurisdictional capture of CORONA. He advised the Air Force Director of CORONA to continue operating as previously, in direct communication with Bissell at the CIA. This amicable judgment did much to enhance spontaneous cooperation between the CORONA effort and "witting" Samos development offices. There was an additional rationale in the basic conviction (of the SAFSP cadre), that CORONA was, at most, an emergency, stop-gap system which would certainly be replaced—and in the very near future—by the sophisticated read-out Samos. In any event, an "ownership" argument over CORONA was considered to be too trivial to be given any attention in Los Angeles or in the Pentagon.

Maj. Gen. Osmund J. Ritland, who had been the first Air Force director of CORONA, was now commander of the newly formed SSD. Ritland had a full understanding of the Samos "problem" and of the need for its streamlined management. As a personal contribution to solving part of the "problem," he recommended that Greer be appointed vice commander of SSD—as an additional duty—thus guaranteeing SAFSP instant access to the Division's talents, resources, and services.

Formation of the National Reconnaissance Office

In August 1961, a year after the relocation of Samos, Charyk forwarded a draft "Memorandum of Understanding," to be signed by Secretary of Defense Robert S. McNamara and DCI Dulles, extending Charyk's responsibilities beyond CORONA and Samos to "all satellite and overflight reconnaissance—overt or covert." This broad franchise was to be called the National Reconnaissance Program (NRP); the managing group would be named the National Reconnaissance Office (NRO). Leadership would be furnished, as additional duties, by the CIA's Deputy Director for Plans (Director, NRO) and the Under Secretary of the Air Force (Deputy Director, NRO). Only the titles were specified; the names of the current incumbents—Air Force and CIA—did not appear. McNamara signed the paper and sent it on. DCI Dulles did not respond.

On 5 September 1961, Charyk sent a second draft of his proposal through DoD/CIA channels. Based on consultation with CIA officials, he designated the CIA Deputy Director for Plans and the Under Secretary of the Air Force as Joint- Directors of the NRO. The following day, Deputy Secretary of Defense Roswell Gilpatric and Lt. Gen. Charles P. Cabell (the Deputy DCI) signed an agreement which:

a. defined the NRP as all satellite and overflight reconnaissance, overt or covert, and

b. established the NRO under the joint leadership of the Under Secretary of the Air Force and the Deputy Director for Plans, CIA.

In a separate action, on the same day, Defense Secretary McNamara designated the Under Secretary of the Air Force as his Assistant for Reconnaissance, with full authority to manage the NRP. But the NSC 5412 Group,[25] reviewing the agreement, withheld approval, questioning the co-director provision.

During this period, important personnel changes were occurring within the Intelligence Community. In November 1961, DCI Dulles resigned from long, honorable service with the CIA. President Kennedy appointed John A. McCone to succeed Dulles—an unusual selection in that the newly elected Democratic President was choosing a Republican as his DCI. McCone was experienced in government; as Under Secretary of the Air Force and, later, as a tough Chairman of the Atomic Energy Commission, he earned the reputation of a battler who usually got his way. Kistiakowsky considered McCone a relentless adversary and, in his memoirs, expressed himself explicitly and profanely on the subject.[26] At the end of February 1962, Richard Bissell—unfortunately the designated victim of the Bay of Pigs fiasco—resigned. He was succeeded, in part, by Dr. Herbert Scoville, Jr., who had been with the Armed Forces Special Weapons Project for six years and with the CIA since 1955; the succession was "in part" because Bissell's Directorate of Plans was to be divided into two organizations: the plans function going to Richard M. Helms and the small technical staff becoming Scoville's (new) Directorate of Research.

**DDR Herbert
SCOVILLE Jr.**

By May 1962, the dust was settling at the CIA, and Pentagon officials reopened negotiations on the reconnaissance-satellite management agreement. On 14 June their discussions culminated in DoD Directive TS 5105.23. This document:

- Established the NRO as an operating agency of the DoD under the direction and supervision of the Secretary of Defense.[27]

- Organized the NRO separately within the DoD, under a Director, NRO, (DNRO), appointed by the Secretary of Defense.

- Made the Director, NRO, responsible for consolidating all DoD satellite and air-vehicle overflight projects for intelligence, geodesy, mapping photography, and electronic signal collection into a single NRP and for complete management and conduct of this program in accordance with policy guidance and decisions of the Secretary of Defense.

On the same date, Deputy Defense Secretary Gilpatric appointed Charyk as DNRO.

The Hazards of Organization

Formal organization of a governmental activity is usually accompanied by an explicit assignment of tasks within that activity. The classic comprehensive analysis of 20th century governmental organization—referred to in a non-pejorative sense as "bureaucracy"—was produced by German sociologist Max Weber.[28] In Weber's steady view, bureaucratic administration develops two contrasting features: (1) a systematic administration characterized by specialization of functions, adherence to fixed rules, and hierarchy of authority, and, (2) a systematic administration marked by officialism, red tape, and proliferation. Weighing the hazards and advantages of bureaucracy, Weber finds them relatively even and observes emphatically that modern government would scarcely survive without the benefits of (1), even though the disadvantages of (2) are a constant, nagging problem.

During four years, the CORONA program had been nurtured to exceptionally successful status, while remaining outside the strictures of "good" or "bad" bureaucracy. In Weber's analysis, such a phenomenon could obtain only under charismatic leadership: "the authority of the extraordinary and personal *gift of grace*" (charisma), which draws followers to it 'in absolutely personal devotion and personal confidence.'[29] This kind of leadership did, indeed, characterize Richard Bissell's presence in the U-2 and CORONA programs. His paucity of engineering expertise was scarcely noticed; in fact, as previously stated, an MIT president had referred to Bissell as "a brilliant project engineer." There had been no need to "regularize" or "bureaucratize" CORONA; in proof, no one ever attempted a CORONA organizational chart or thought of specifying its "owner."

CORONA's CIA and Air Force units had chosen to remain very small and very busy. In the CIA CORONA Office, the majority of key persons had been drawn from the Air Force, either as active duty designees or retired officers choosing a second career. These people were specialists in aircraft operations, mission planning, photographic equipment, and aeronautical engineering and were furnished freely and cheerfully in the spirit of Air Force-CIA partnership. (Interestingly, these Air Force officers, to a man, strongly opposed even the suggestion of change in managerial "structure.")

Charyk's leadership qualities were as exceptional as Bissell's and equally well-known and appreciated in high places. When Kistiakowsky was searching for the best environment in which to place Samos, Charyk had convinced him to move the project to the Office of the Secretary of the Air Force (Kistiakowsky then "sold" this idea to a President who had previously declared that only the Office of the Secretary of Defense could be trusted with high-risk, high-priority development programs). Secretary McNamara and Deputy Secretary Gilpatric were similarly impressed by Charyk and trusted him implicitly for advice, counsel, and technical judgment.

Now that all overhead-reconnaissance developments and operations had been designated to the NRO and Bissell had departed, another Weber "law" would begin to apply: "the routinization of charisma,"[30] in which "the [initial] genuine charismatic situation quickly gives way to incipient institutions."[31] Predictably, there would soon be drafted an internal "structure" which would list the extent of responsibility of each operating subunit of the NRO. This "structure" would inevitably be supported by fixed rules, explicit functional duties, and a careful definition of jurisdictional areas.[32]

The Tyranny of Organizational Charts

Formalization of the NRO organizational process began with a "picture"—a chart—showing all of the newly assigned assets. These were (1) the CIA's overflight aircraft—U-2s and A-12s; (2) the Navy's POPPY satellite (an electronic intelligent [elint]) collector, directed toward frequencies used by Soviet naval radars); (3) the CORONA photo-satellite; and (4) a family of Samos satellites in various stages of development. The NRO Staff Director Col. John L. Martin, Jr. had been told to sketch this picture; his first draft showed this arrangement:

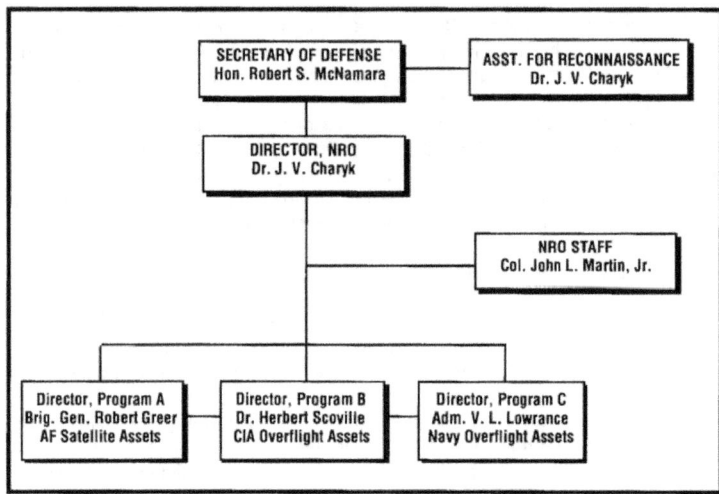

Based on long experience, Col. Martin was sensitive to the dangers implicit in bureaucratic structure and was determined to delay or prevent them. It was his hope that the NRO could be developed into a loose *confederation* of activities, bound together by the diplomatic skill of its Director. His initial version of an organizational sketch showed an ingenuous evasion: it cautiously skirted the question, "Where does one slot CORONA?" But CIA's Director of Program B would be sure to ask why Greer's box was the only one to use the word "satellite."

And there was another problem. Col. Leo P. Geary, the Air Force Staff contact for the U-2 and A-12, had observed, rather vehemently, that he should be represented as a *Director* on the NRO chart, in parallel with Programs A, B, and C. When it was suggested that his function was, at most, a *staff* function and that he might, perhaps, be listed as a member of the NRO staff, he reacted even more strongly, appealing his case to the DNRO and the AF Chief of Staff.

Although Charyk considered the staff officer appellation a reasonably accurate description of Geary's duties, he was also recalling long-drawn-out negotiations in developing the basic NRO agreement. He decided to "absorb" Geary"s grievance, rather than invite further argument and instructed Col. Martin to create a Program D, with aircraft and drones under Geary's aegis. Whereupon, a new chart appeared:

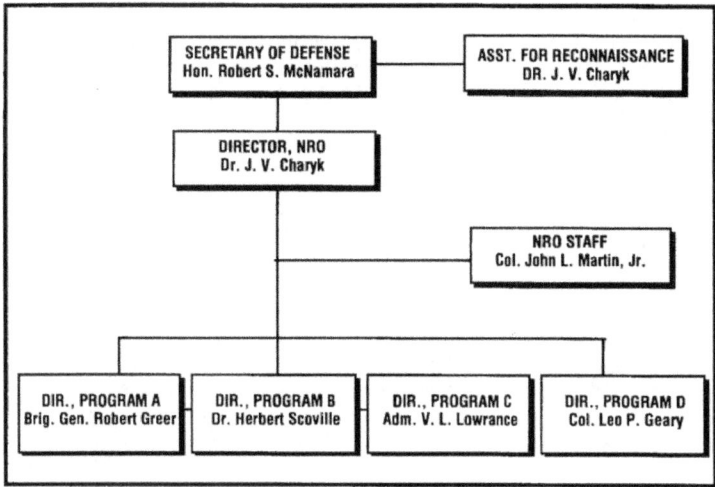

The chart was new, but an old question remained: Did the chart say anything important? What happened when one positioned overflight assets within these austere boxes? With regard to Program A, Greer certainly "had" Samos; Lowrance, in Program C, was building POPPY; Geary's Program D definitely assisted the Strategic Air Command (and the CIA) in operating drones and overflight aircraft; but, now the hard question, "What are the 'holdings' within Program B?"

The only unassigned residual was CORONA—the organizational chart trumpeted that fact by omission. In happier day's, the location of CORONA management authority had never been defined or even questioned; it hovered somewhere between Los Angeles and Virginia in a nebulous Valhalla; to identify it, one would have had to assign it—and that would have served no useful purpose to the Air Force or to Bissell. But Bissell was gone now, and, in January 1963, it was learned that DNRO Charyk—one of the very few persons who could have nurtured the organization through an awkward era—was planning to leave, to become president of the Communication Satellite Corporation, Comsat.[33] With Charyk and Bissell gone, the era of charismatic leadership was coming to an end, and the NRO would be threatened by the danger of moving toward the darker side of bureaucracy (tagged by Weber as "officialism and proliferation").

The Navy's Program C and Air Force's Program D would remain serene, skillfully carrying out existing, well-defined roles. Unhappily the Air Force's Program A and CIA's Program B would soon enmesh themselves in endless petty arguments over CORONA functions, responsibilities, and prerogatives. To newcomers in Los Angeles, it would appear that the Air Force had been doing practically all the work on CORONA and, therefore, had the right to make unilateral decisions regarding the program's future. At Langley, Virginia, newcomers would be told that the CIA, in 1958, had rescued CORONA from oblivion and had singlehandedly achieved success, more than earning proprietary "rights" to the system. The situation would be aggravated further by condescension on the part of Samos personnel toward the CORONA program—they described CORONA as a lash-up, a temporary expedient which would be replaced, very soon, by the Samos family of satellites. Newcomers to SAFSP would ask why all the fuss about CORONA ownership; the future would certainly belong to Samos.[34]

Scoville had an additional problem—totally unknown to Program A, the NRO staff, or the DNRO—which affected his outlook profoundly. He believed he had a high-level, external mandate (and he did, see Section 6) to strengthen the technological capabilities of the CIA. At present, that capability was very thin, largely dependent on the (now-habitual) practice of borrowing technical specialists from the military services. Scoville hoped to change all that, but, instead of being supported in his efforts, he was (1) being denied the manpower "billets" which he had "lost" in the dissolution of the Directorate of Plan's Office, (2) receiving negligible CIA financial support, and (3) now threatened by the prospect of losing even his small (CORONA) holdings to the Air Force. A profound pessimism began to affect Scoville's outlook and personal relationships. An NRO Staff Director described the change: "When 'Pete' [Scoville] began working with the NRO, he used to visit the NRO's Director and request concurrence on new plans or actions. We knew that things had changed when 'Pete' began to go to McCone *first*, and then drop over to *tell* the DNRO what he and the DCI had decided to do."[35] But even this operating mode did not reassure Scoville; he left the CIA in June 1963. His replacement was Dr. Albert D. Wheelon, the CIA's Director of Scientific Intelligence and former missile expert with the Ramo-Woolridge Corporation.

An Open "Futures" Function

As months passed, it became increasingly difficult—and dangerous—to develop explicit functional statements for Program A (Air Force) and Program B (CIA). But this condition, which would have been very distressing to a normal "seasoned bureaucrat," had an unanticipated wholesome effect upon the NRO: it inhibited transition from "good" to "bad" bureaucracy.

DDS&T Albert WHEELON

A seasoned bureaucrat, examining the NRO organization, would have (1) deplored the absence of defined "turf" and (2) pointed to a "fatal" weakness: "You have not provided a central office for planning *follow-on* reconnaissance systems!" He would then have cited the advantages of unified planning. "Place one Central Planning Office on your NRO Staff, where it can serve the future needs of *all* your programs. Such an ecumenical staff office will be able to draw on the expertise of all program offices and will represent the entire community need for follow-on systems." (To which the ghost of Weber would have replied, "Do that, and you will dig the grave of NRO planning. Do that, and you will hasten the advent of bad bureaucracy.")

The strength of the future NRO—the hope for giant strides in improved overhead reconnaissance systems—lay in *good* bureaucracy. And good bureaucracy demanded that future system planning remain an open, competitive organizational function, fully receptive to the best ideas and efforts of each Program Office and sufficiently mature to endure the hazards and even the possible battle damage of aggressive competition. One had to keep room in the NRO for "What Ifs." What if CORONA was not a *short*-term expedient? What if the Samos "family" was not to become the follow-on to CORONA? What if US Intelligence Board (USIB)[16] requirements shifted? What if a second strong satellite development office began to emerge? What if new charismatic figures appeared on stage? What if . . . ?

As long as the NRO never quite managed to get organized, as long as the "futures function" remained open, it could be hoped that its planning would be strongly competitive, in the spirit of free enterprise. One might even expect such an environment to enhance occasional charismatic renewal.

Section 6

A New Space Claimant: FULCRUM

Since the days of the Eisenhower presidency, the CIA had been under continuous pressure to improve its scientific and technological capability to collect and evaluate intelligence information. The pressure began with Dr. James R. Killian, Jr., who headed Eisenhower's TCP in the mid-1950s; it was repeated by Dr. Edwin H. Land, president of the Polaroid Company and long-time presidential advisor. Neither DCI Dulles nor his successor, McCone, had done much about these recommendations and, as indicated earlier, Scoville had resigned over the Agency's failure to form an effective scientific directorate.[37] Scoville took his action in spite of Headquarters Notice 1-9, 16 February 1962, which established the Office of the Deputy Director for Research (effective 19 February 1962); he was convinced that there was no immediate prospect of acquiring the resources needed by such a directorate.

The Advent of Wheelon

In order to attract Dr. Albert D. "Bud" Wheelon as a replacement for Scoville, it was necessary for the DCI to guarantee the people and authority needed to build a strong technological capability. On 5 August 1963, Wheelon did become the CIA Agency's first Deputy Director for Science and Technology (DDS&T).[38] He saw his primary need to be carefully selected, highly skilled people and soon began to recruit them. Despite temporary problems and bickering with Program A (Air Force) over the CORONA Program, the DDS&T Staff soon shifted its attention to two truly ambitious efforts: (1) the creation of a new search *and* surveillance system and (2) the initiation of ▬▬▬▬▬▬▬▬▬▬▬▬▬▬▬▬▬▬▬▬▬▬▬▬▬▬▬▬

With the departure of Bissell and Charyk, the NRO organizational center of gravity had shifted. The new DNRO, Dr. Brockway McMillan, found himself vis-a-vis a hard-driving *competitor*: DCI John McCone. McMillan had lived the patterned, reasoned life of Bell Laboratories; McCone knew the jungle law of heavy-gauge infighting in Washington's corridors and had a long record of success in getting what he wanted (including Livermore Laboratories). Even before Wheelon became DDS&T, McCone had declared that something had to be done "to get the CIA back into the satellite business, including developing proposals for a new and better system beyond CORONA."[39]

In May 1963, McCone convened a Scientific Advisory Panel under the chairmanship of Dr. Edwin Purcell, Nobel laureate and professor of physics at Harvard University, "to determine the future role and posture of the United States Reconnaissance Program," an undertaking which one would have expected to be functionally within the purview of the DNRO.[40] The following month, this Panel recommended a CORONA improvement program for optimizing system performance. Neither this nor

Edwin H. LAND

DCI John A. McCONE

NRO Director Brockway McMILLAN

subsequent studies went beyond evolutionary improvement of CORONA until Wheelon tasked the National Photographic Interpretation Center (NPIC) to determine what photographic resolutions were needed to identify a wide variety of Soviet targets. The study, made by 25 NPIC photointerpreters, was completed in January 1964; it concluded that the majority of Soviet targets could be identified with photographic resolutions of 2 to 4 feet.[41] At a time when CORONA was acquiring 7 to 10 foot resolution, NPIC's finding was a strong testimonial to the need for a new search and surveillance system. It played back what was the basis of the study, namely, the utility of a system with GAMBIT resolution and CORONA coverage.

The result of Wheelon's NPIC Study could not have been a surprise to DNRO McMillan.[42] In a 12 December 1963 note to Defense Secretary McNamara—on some NRO/CIA issues (and there were now many)—McMillan had suggested that "the final price of peace with the CIA 'considering the temperament of its leaders' was at least to give the CIA *carte blanche* for development of a new search system." He stated that until something of this sort was done, or the CIA leadership changed, there would be continual obstruction to the NRO and its actions.[43]

In February 1964, as an augmentation to its own in-house study effort, Wheelon's office contracted with Itek Corporation "to determine the feasibility and potential intelligence value of various sensors in satellites."[44] Itek confirmed the results of the NPIC study. In April 1964, the CIA directed Space Technology Laboratories (STL), of the Thompson-Ramo-Wooldridge (TRW) Corporation, to investigate a spinning vehicle hybrid system. The study funds were, of course, provided by the NRO.

The Advent of FULCRUM

CIA documents state that in May 1964 (three months after the initiation of the DDS&T-sponsored work) "each effort, the Agency's as well as Itek's and STL's, independently concluded that we needed CORONA-type coverage with consistent GAMBIT-type resolution."[45] On this basis, Wheelon reportedly proposed to the DNRO a system codenamed FULCRUM with:

- A 5,500-pound photographic payload, using a Titan-II booster

- Two 60-inch focal length stereo cameras with nadir ground resolution of 2 to 4 feet over a strip 360 miles wide

- 68,000 feet of 7-inch-wide film covering 11 million square miles for each mission (and requiring a new reentry vehicle)

- An estimated cost of ▮▮▮▮▮ per launching.

There is no mention in the NRO Chronology of this proposal; but there is record of a complaint on 12 June 1964, by McMillan to Deputy DCI (DDCI) Gen. Marshall S. Carter that McMillan had "separate indications that Dr. Wheelon was contracting for satellite system and subsystem studies with . . . instructions to the contractors 'not to give the DNRO or DDR&E[4b] [Director of Defense Research and Engineering] any information regarding the source of the request for study.'"[47]

DNRO McMillan, together with DCI McCone, DDCI Carter, DDR&E Dr. Eugene Fubini, and others, was officially informed of the FULCRUM concept on 15 June 1964 and was asked by McCone to review the proposal with his technical personnel, including Greer from SAFSP.[48] On 25 June 1964, McMillan recorded a discussion with DDCI Carter in which several agreements were made relative to the conduct of further FULCRUM studies.[49] Among these was the statement that "CIA funds to the extent of about ▮▮▮▮▮▮ could be obligated in FY64 to conduct tests at Itek of an engineering model of the critical film-transport mechanism." It was further agreed that "in the event the FULCRUM concept was approved for development other than under CIA auspices, the CIA funds expended on the tests would be reimbursed by the NRO." (The existence of this "money-back guarantee" implied a lack of certainty on the part of both McMillan and Carter, at that time, as to whether CIA would ultimately manage the program.) This was followed by McCone's statement, on the following day, that "the DNRO should be directed to establish FULCRUM as an NRO development project, and should assign responsibility for research, development, and operation" to the CIA.[50]

Eugene G.
FUBINI

Lt. Gen. Marshall S.
CARTER

To ensure that all bases were covered, the DCI asked Dr. Land to convene a panel "to consider the technical feasibility of a newly proposed satellite photographic system called Project FULCRUM."[51] In addition to, Land, the panel consisted of Dr. Allen F. Donovan; Dr. Sidney D. Drell; Dr. Richard L. Garvin; Mr. Spurgeon W. Keeny, Jr.; Dr. Donald P. Ling; Mr. Arthur C. Lundahl; and Dr. Aden B. Meinel. This group met on 26 June 1964 and, after "a day-long presentation on FULCRUM by representatives of the DDS&T and selected contractors, held an executive session and prepared recommendations to the Director."[52] (Mr. Lundahl, head of the NPIC and a CIA employee, excused himself from participation in the panel's recommendations.) In an oral report to McCone, Dr. Land called the proposed system "extremely attractive" and "praised the ingenuity of the idea."[53]

It can be assumed that cited instructions from DDS&T Wheelon relative to the DDR&E ("not to give the DNRO or DDR&E any information . . .") were in anticipation of a negative reaction from those offices. A 30 June 1964 memorandum from Fubini to McMillan referred to the FULCRUM briefing and offered the following summary:

- "The Purcell Committee advised against a new broad coverage system.

- The Air Force made a series of recommendations for the improvement of the CORONA camera, in accordance with Purcell Committee recommendations.

- Dr. Wheelon disagreed with the Air Force recommendations and sponsored the Drell Committee study.

- The Drell Committee found little correlation between the product results and the mechanical or optical characteristics of the system and made a number of suggestions for further quantitative measurements of the product.

- Recent CORONA missions seemed to confirm the Purcell [Committee] recommendation that substantial improvement over the CORONA camera result could be obtained and appeared also to confirm the Drell Committee findings, since there did not appear to be any basic change in the camera setup between recent missions and previous ones.

- The CIA made a proposal called FULCRUM, which did not correct the unknown[54] defects of the CORONA camera or take into account the questions, recommendations, or conclusions of the Drell Committee relative to hardware improvement, but, instead, proposed to initiate a completely different camera design.

- Recent results in CORONA 'take' seemed to indicate a possible resolution of 5 to 7 feet, in rough accordance with expectations. If this resolution were maintainable, would there be sufficient motivation for a new broad coverage system in the 3.5- to 5-foot resolution range? (CIA studies seemed to indicate that resolutions substantially better than this value were desirable for high target-detection confidence in many target classes.)"

~~SECRET~~
~~NOFORN-ORCON~~

Fubini stated that he considered it absolutely necessary, before a new system design were accepted, to compare the old CORONA results, the new CORONA results, the Drell Committee results, GAMBIT[55] results, and, finally, the technical recommendations for the new broad-coverage camera, to ensure that the (still unknown) causes of poor performance in CORONA had been eliminated. Fubini also expressed his belief that a substantial amount of effort could and should be devoted to these problems at the earliest possible time.[56]

Despite these cautionary views, on 2 July 1964 (only three days after the Fubini memorandum) Wheelon presented a plan to the DNRO for initiating FULCRUM.[57] Wheelon's plan called for:

- ▓▓▓▓ funding for a six-month design analysis by seven contractors for a new camera system, a new reentry vehicle, and a new spacecraft; launching to be by a Titan-II booster from the Pacific Missile Range (PMR). STL of the TRW Corporation had been chosen as the integration, assembly, and checkout contractor.[58]

- Establishing, under CIA's DDS&T, a FULCRUM Project Office with ▓▓▓▓ technical people (most of whom would be new hires) to perform system engineering and technical direction.

- Enlarging his project staff (by further recruitment) to approximately ▓ people.

- Providing procurement/contracting and security for FULCRUM.

- The DoD to provide launching and capsule-recovery services, beginning in FY67.

The reaction to Wheelon's proposal came quickly. In an 8 July 1964 letter to DCI McCone, Deputy Secretary of Defense Cyrus Vance referred to the CIA plan and suggested that "in order to insure that all possible alternatives have been explored . . . we should ask Director, NRO, to direct the completion of comparative studies, meanwhile authorizing CIA to pursue only those designs and tests that are necessary to establish the feasibility of the proposed FULCRUM camera concept."[59] Vance expected that the results of other studies would be available in six months (by January 1965); this would allow a determination as to whether a new system should be developed, facilitate selection of the system to be developed, and provide a basis for assigning responsibility for system development and operational employment.

Wheelon also responded quickly, on 10 July 1964, with a memorandum that not only confirmed his earlier request for ▓▓▓▓ of NRP FY65 funding for the six-month design analysis effort (to which he would add ▓▓▓▓ of CIA funds), but further asked that "the remainder of the ▓▓▓▓ sought in FY65 be set aside for Program B use, pending the outcome of the initial tasks scheduled for a period of six months."[60] That Wheelon's plans for FULCRUM went well beyond "comparative studies" is clear from a summary of the program which was attached to a memorandum from Wheelon to the DDCI, dated 23 June 1964.[61] In that document, a funding requirement of between ▓▓▓▓ for FY65 thru FY69 is summarized.

~~SECRET~~
Handle via
BYEMAN-TALENT-KEYHOLE
Control Systems Jointly
BYE 140003-92

Also on 10 July 1964, Jackson D. Maxey was named FULCRUM Project Manager. Maxey was one of the senior engineers hired from industry by Wheelon, using a separate, higher pay scale for scientists and engineers that had been established when the DDS&T was organized. Maxey was Chief of DDS&T's ▓▓▓▓▓ ▓▓▓▓▓▓ and was supported by a project engineer, Leslie Dirks (another recent hire), and an executive/administrative officer, John N. McMahon. The quality of the FULCRUM staff was demonstrated by the fact that Dirks, the "father" of the ▓▓▓▓ new-real-time system, later became the CIA's DDST, and McMahon ended his CIA career as DDCI. Clearly, McCone and Wheelon were very serious about building a strong space system development and management capability.[62]

Deputy Secretary of Defense Cyrus
VANCE

Leslie C.
DIRKS

Ratification of the need for a new reconnaissance system was recorded by the USIB on 27 July 1964; it approved, as guidance to the NRO, the recommendation of its Committee on Overhead Reconnaissance (COMOR) that there was a need for a search and surveillance system capable of CORONA coverage and GAMBIT resolution.[63] This echoed the CIA justification for FULCRUM, which had been presented as a system to replace both CORONA and GAMBIT (with concomitant reduction in total costs).

Vance Sets Limits on FULCRUM

An important, but somewhat limiting, step was taken as a result of a 29 July 1964 letter from Deputy Defense Secretary Vance to DCI McCone[64] and a subsequent meeting on 11 August 1964 attended by Vance, McCone, Fubini, and McMillan. In the 29 July letter, Vance had iterated the agreed-upon objective of FULCRUM: to establish, in an expeditious manner, definitive data on the technical issues critical to the performance or success of the camera. Vance stated his belief that the FULCRUM effort should be directed toward and limited to:

1. Initial design, fabrication of an engineering model, and definitive testing of the complete film-transport mechanism.

2. Preliminary optical and mechanical design of the rotating camera, limited to the amount necessary to establish a model suitably simulating the camera's mass inertia, balance, and flexural stiffness; this model should be dynamically tested with prototype bearings.

Vance further stated that activities should be conducted under the following general conditions:

- Under the aegis of the NRO, with full information on activities and progress made available to the NRO at all times.

- Separate contracts for items 1 and 2, above.

- Consideration of competitive bidding on item 1, above.

- Application of funds only to specific contracts, each defined by a negotiated statement of work approved by the NRO and accompanied by a definitive contractor cost estimate.

- No contracts for items not covered in items 1 and 2 above (that is no contracts for system integration, spacecraft design, reentry vehicle design, and so forth).

- An individual in the CIA to be identified as responsible for the contract.

Vance provided several additional minor suggestions relative to the activities and requested McCone's comments.[65]

McCone expressed his general agreement on 11 August 1964.[66] It is recorded that "Mr. McCone stated that it was not his intention to establish within the CIA a unilateral capability for development and operation of space systems."[67] He believed that responsibility for launching and on-orbit operation of systems would remain with the Air Force. It was also agreed that should a FULCRUM development be undertaken, the CIA would not do system engineering in-house, but would rely on a contractor for that function.

In addition to the camera work described by Vance, it was agreed that a system design study would be undertaken on FULCRUM. These terms were specified:

- The study would be conducted by a contractor, or contractors, and limited to about a six-month period.

- No commitment to a subsequent development would be made.

- Should a development be undertaken, contractors would again be selected by competition and the study contractor(s) would have an opportunity to bid.

- If feasible, study contractor(s) would be selected competitively.

- The study would be under the aegis of the NRO and NRO funds would be authorized against firm negotiated proposals.

The 11 August meeting was followed two days later by another meeting attended by McMillan, Wheelon, Brig. Gen. James T. Stewart (Director of the NRO Staff), Maxey, McMahon, and Col. Strand (McMillan's military aide) to discuss the scope of Phase-I activity in Project FULCRUM. McMillan saw Phase-I as "a period of system design study; that in addition to study efforts regarding camera design and fast film transport, should also consider the housing for the payload such as the spacecraft." He suggested that "the Titan-III married to an Agena" be considered and went on to state "that the National Reconnaissance Program was ripe for a new recovery vehicle and possibly two." He acknowledged "that the FULCRUM R/V requirements were far more demanding than anything we now have." During the meeting McMillan questioned the CIA's role in system engineering and technical direction, an issue which was not resolved for two months.[68]

McCone Broadens the Limits on FULCRUM

It is clear from a 14 August McCone memorandum[69] that thus far Wheelon had only a limited mandate in FULCRUM. In the memorandum, McCone said he would "make two points abundantly clear" with regard to the handling of FULCRUM contracts:

1. "There shall be no commitment, contractual or implied, that we are to proceed past the authorized research and development (R&D) work on the film-handling mechanism and the camera, which includes developmental mockups built in sufficient detail to answer or to disprove all questions or doubts concerning feasibility and, with respect to the spacecraft and reentry vehicle, conceptual designs and sufficient detailed engineering to present accurate determinations as to weight of the total assembly and compatibility with the launcher.

2. You will employ engineers and contractors to the fullest possible extent, reserving as 'in-house activities' responsibilities for supervision and guidance of the engineers and contractors. I wish you to avoid as far as possible unnecessarily building an in-house capability, restricting the expansion of your staff, if any is required, to such additions as are necessary to adequately supervise the work of the engineers and the contractors."

McCone went on to state that this guidance specifically indicated:

- Employment of an architect-engineer or system engineering contractor to be responsible for developing plans, specifications, etc., for all phases of the project.

- Competitive contracts with two or more contractors for the film-transport mechanism.

- A contract for the camera, recognizing that it probably could not be competitive because of the Itek input to the FULCRUM concept.

- Competitive contracts for the design of the spacecraft, assuming that competitors would introduce first-phase conceptual plans, from which the winning contractor would be chosen and authorized to proceed with detailed engineering.

While McCone's direction to Wheelon was somewhat limiting, it still went beyond the DCI's agreement with Vance. The Vance letter had limited current FULCRUM efforts to design, fabrication, and testing of the film-transport mechanism and preliminary optical and mechanical design of the camera; it precluded contracting for anything beyond that activity. In addition, it specifically precluded "system integration, spacecraft design, r/v design, etc." Thus, only three days after his 11 August meeting with Vance, where he had agreed to Vance's plan, McCone was telling Wheelon that he was authorized to proceed with items precluded by the Vance plan. (McCone's direction is not inconsistent with McMillan's views on the scope of the Phase-I effort, as reflected in his comments during his meeting with Wheelon on 13 August 1964.) McCone's letter to Wheelon is interesting from another point of view in that he is seemingly authorizing a staff buildup only for the purpose of allowing Wheelon to "adequately supervise the work of the engineers and the contractors." From this wording it would appear that the "engineers" referred to were not to be people of Wheelon's organization. This despite McCone's clear approval of Wheelon's unique pay scale for scientists and engineers, a factor which allowed Wheelon to build an in-house technical capability of very high quality.

Internal CIA correspondence then circulated, allowing the FULCRUM effort to proceed. In a 27 August 1964 memorandum[70] to the DDS&T, DDCI Lt. Gen. Marshall S. Carter provided additional DCI-approved guidelines for organization and direction of the FULCRUM program. On 31 August, Wheelon responded with an internal CIA plan and terms of reference;[71] these were approved by McCone and Carter on 1 September.

Vance-McCone and System Engineering/Technical Direction (SE/TD)

Although Wheelon's letter of 31 August was purportedly both terms of reference and a program plan, what it really amounted to was a brief history, mention of the principal tasks to be accomplished, and an idea of who would be tasked to do what in the near future. The actions envisaged were consistent with early instructions from both Vance and McCone on FULCRUM. Despite this fact, there were persistent areas of disagreement as to what was to be done by the CIA on FULCRUM.

In a 4 September letter to McCone, Vance called attention to the fact that the FULCRUM program direction issued by DDCI Carter on 27 August did not reflect the Vance-McCone agreement in one very important detail.[72] The area of contention was the planned role of the FULCRUM system engineering contractor. Vance now added something to the previously identified system engineering contractor's title, making him the SE/TD contractor. Adding the technical direction role to the contractor's responsibility meant to Vance that CIA employees would not provide technical direction to FULCRUM.

The usage "SE/TD" had come into being over a decade earlier in the Air Force's ballistic missile program. The Air Force had originally charged the Ramo-Woolridge Company with an SE/TD function for that program. It should be noted that while R/W did indeed do (and now, as TRW Corporation, still does) system engineering for the ballistic missile program, its technical direction function was not a clear untrammeled activity. The problem was simply that the government could not devise a contractual procedure for allowing one contractor (the SE/TD contractor) to direct the technical affairs of another contractor (the "performing contractor") whose contract was with the same government activity as the SE/TD contractor's. It was easy enough if the performing contractor was a subcontractor to the contractor responsible for SE/TD; the problem arose when both held prime contracts with the government. Most technical direction involves changing, in some form, the scope of effort the performing contractor is undertaking. Such a "change in scope" inevitably brings the government into the process. It is not recorded why Vance, in face of the Defense Department's operational experience with difficulties inherent in "technical direction," chose to take such a strong position on having a contractor, as opposed to CIA people, perform that direction on FULCRUM.

It can be surmised that staff members of the Office of the Secretary of Defense—who had not faced the realities of operating an SE/TD contract—may have suggested the approach to Vance in order to forestall a buildup of technical management capability in the CIA's fledgling DS&T, possibly seeing it as either competitive with, or redundant to, existing management assets of the Air Force.

The issue was not quickly or easily resolved. It was discussed, without conclusions by Vance, McCone, McMillan, Carter, and Fubini in a 14 October meeting. Neither was it resolved in a 21 October telephone conversation between McMillan and Wheelon. In response to McMillan's question as to "whether the DCI had made any determination about incorporating technical direction language into the FULCRUM systems engineering contract . . . Wheelon stated that if the NRO had the impression the DCI was considering such a move, it was mistaken. Dr. Wheelon indicated that CIA had absolutely no intention of incorporating technical direction in the way he and Dr. McMillan understood the term."[73] The available record indicates that at a subsequent NRO Executive Committee[74] budget session it was stated that "McCone would review the contractual language defining the SE/TD role of the Aerospace [Corporation] on GAMBIT" to see if he considered that approach appropriate for FULCRUM. It should be noted that Aerospace Corporation had system engineering responsibility, but no technical direction role, in that program. Whether McCone made such a review is not recorded; however, technical direction remained a CIA, not a contractor, responsibility.

Meanwhile, work on FULCRUM was proceeding. Wheelon asked McMillan to keep him informed on current and planned reentry vehicles "so that we do not design two capsules where one might be justified;" he also informed McMillan that, at McCone's direction, he was looking at both the Titan-II and the planned Titan-III booster systems for FULCRUM (and other applications) and requested additional Titan-III data.[75]

In September, the CIA began actions which resulted in the competitive selection of General Electric (GE) as the spacecraft contractor and Avco as the reentry vehicle contractor for the Phase-I FULCRUM study (which began in September 1964 and would end on 31 January 1965). The planned funding for Phase-I was ███████ and was to be followed by Phase-II (development, production, and operation of the system), which was to begin on 1 March 1965.[76]

On 1 September, Wheelon, "with the knowledge and concurrence of the DCI . . . , created a Special Projects Staff (SPS), as an interim mechanism for managing the CIA's NRP activities. The personnel ceiling and incumbents of the Systems Analysis Staff of the "S&T" were made available to SPS, and Mr. Jackson Maxey was named Chief of this temporary management staff." (Maxey had headed the Systems Analysis Staff). SPS, as a formal organization, did not come into being until early 1965.[77]

In early December 1964, concern over the validity of booster costs and the availability of boosters led Maxey, John Crowley,[78] McMahon, and Richard DeLauer (of STL/TRW) to visit the Martin Company plant in Denver, Colorado, where Titan vehicles were produced. They concluded that "no technical bottlenecks existed in supplying missiles" nor "in getting adequate resources via Martin/Denver to run a completely civilianized launching facility."[79] The "civilianized launching facility" concept would have required the CIA to contract directly with Martin, for not only the booster but for all launching services up to injection into orbit. Martin preferred this approach and noted that selecting it should save about 20 percent as compared to purchasing through the Air Force. Such an arrangement was not, however, consummated.

Section 7

Competing Claimants: FULCRUM and S-2

In early 1964, before the CIA got under way on FULCRUM, the DNRO had authorized SAFSP to begin two separate efforts for formulating a concept and preliminary design of the photographic payloads for an optimal search and broad-coverage satellite system. These efforts had been given the designator S-2. At Eastman Kodak, S-2 work had begun in the fall of 1963, when SAFSP redirected Eastman Kodak's work on VALLEY.[80] At Itek, S-2 work did not begin until 18 November 1963. Both Kodak and Itek had completed S-2 preliminary designs by September 1964, just when the CIA was starting its Phase-I FULCRUM program. The same month, SAFSP broadened its S-2 efforts, offering similar contracts to Fairchild Camera and Instrument Company and to the Perkin-Elmer Company. Perkin Elmer declined, but Fairchild began "a five-month design study which produced a design concept which pushed the state of the art in refractive optics."[81] In December 1964, the SAFSP Advanced Development Project Office, under Col. Paul Heran, "initiated competitive parametric studies of a possible orbiting vehicle at both Lockheed and General Electric, and . . . began investigation of booster requirements,"[82] in support of S-2.

In the early 1960s, intercommunication among NRO Program Offices was not very effective. As an example, it was not until August 1964 that the "CIA received informal word that Dr. McMillan (through SAFSP, Maj. Gen. Greer) had started efforts in competition with FULCRUM on behalf of the Air Force at Eastman, Fairchild, and Itek.[83] As previously noted, the Eastman Kodak and Itek efforts had been going on for 10 months or more; the Fairchild effort was contractual a month after the "informal word." Wheelon, in reporting this to DDCI Carter, took the erroneous[84] view that CIA efforts on FULCRUM had stimulated competitive studies within the Air Force; he opined that it was "shameful to learn about it from private industry."[85] It should be noted that, as a result of the Land Panel review of FULCRUM on 25 June 1964, Wheelon had been aware that the Air Force's VALLEY program "was designed to accomplish the same result as FULCRUM, but in a different manner."[86]

There were persistent differences of opinion as to what the CIA had been authorized to do on FULCRUM. In a 29 September 1964 memorandum to Wheelon, McMillan noted that he had been advised that the CIA had initiated funded spacecraft and recovery vehicle competitions. McMillan considered these premature and not in conformance with the 11 August NRP ExCom agreements; he requested suspension of further efforts until the situation had been considered by the ExCom. In McMillan's view, all that the 11 August agreement permitted was "in addition to preliminary design in the FULCRUM camera, and design and test of the film transport system, a contractor should be engaged to conduct a comprehensive systems design study centered on the FULCRUM concept."[87] Weelon responded that the CIA's plans, which included the spacecraft and reentry vehicle efforts, were those agreed to in a meeting attended by Vance, McCone, Fubini, and Eugene Kiefer, Deputy Director of the NRO. He said that at this meeting McCone had included efforts beyond those cited by McMillan (in his 29 September memorandam to Wheelon) and the group had agreed with McCone's presentation.[88]

The Vance-McMillan Task Force and Steering Group

With the completion of task 1 of the FULCRUM program in sight and the completion of the payload preliminary designs of S-2 accomplished, it seemed an appropriate time for Vance to propose that McMillan set up a task force, guided by a steering group, "to assure that the approach or approaches selected for future development of a new search and/or surveillance system fulfilled *all* national requirements and were, in fact, the best options available."[89] Vance told McCone of his intention on 19 November 1964 and said "that he had asked the group to examine information needs, determine technical and operational criteria, and present an evaluation of the most promising alternative search and/or surveillance satellite systems which might be included in the NRP."[90] He envisaged the task force as operating in the Washington area on essentially a full-time basis. Vance asked McCone to provide a CIA representative to both the task force and its parent steering group. On 8 December, the CIA designated two representatives to the steering group: Gen. Carter nominated Huntington Sheldon as the initial CIA representative; he also named Arthur Lundahl, Director of the NPIC. (Sheldon was replaced by Dr. Max S. Oldham on 14 December.)

Despite CIA "participation" in McMillian's task force and steering group i soon became evident that the Agency had misgivings about the focus and purpose of the activity. On 25 November, McMillan asked Wheelon to furnish a FULCRUM briefing on 9 December to "the steering group for the new NRO Search/Surveillance Satellite System."[91] On 30 November, Wheelon responded that "he would have to await instructions from 'his boss' before agreeing to brief the steering group as requested" and added that "his organization was not persuaded that the steering group was a proper or good idea."[92]

In a discussion with McMillan in early December, DDCI Carter referred to McMillan's request that the steering group be briefed on FULCRUM on 9 December 1964 and "advised that Mr. McCone's letter to Secretary Vance had excluded FULCRUM from the consideration of the steering group" and that "he would discuss the matter with Mr. McCone as the first order of business after his [McCone's] return."[93] In a 14 December 1964 memorandum for McMillan, Carter pointed out that participation by CIA people in the work of the steering group and task force did not in any way commit the DCI or the CIA to the findings of these groups, specifying that they were participating as individuals who had the technical competence needed "in Dr. McMillan's studies" and that "substantive actions developed as the result of studies. . . would be subject to the approval of the DCI and, as appropriate, the USIB."[94]

The Land Panel and FULCRUM

It should be noted that despite the painstaking establishment of a steering group and task force at the behest of the Deputy Secretary of Defense, there is no evidence that these activities accomplished their assigned functions. It turned out that the CIA actually performed the basic system evaluation, using one of its high-level technical advisory groups, headed by Dr. Edwin Land. In July 1964, on McCone's initiative, the

Land Panel had independently evaluated the FULCRUM concept and had recommended a six-month feasibility study, which was now nearing completion. In early February, DDCI Carter, acting for McCone, informed "the people in the Pentagon that he was going to convene a panel of technical experts, and that before February was over he expected that FULCRUM would either be cancelled or going as a full-scale development effort."[95] On 16 February, Carter stated that "Land had agreed to reconvene his panel to evaluate the results of the [FULCRUM] feasibility program, but that CIA did not feel that it should include government people."[96] In a meeting attended by Land, Wheelon, McMillan, and Fubini on 16 February 1965, the following were chosen to serve on the Panel: Dr. Edwin Land, chairman; Dr. Sidney Drell; Dr. Donald Ling; Dr. James Baker; Dr. Allen Puckett; Dr. Edwin Purcell; and Dr. Joseph Shea. (Dr. James Killian and Dr. William Duke were also named but were unable to serve.)[97]

As a prelude to this critically important evaluation, a briefing on the status of FULCRUM was given at Itek on 18 January 1965. Attendance was large: the CIA was represented in addition to Land, by McCone, Wheelon, Maxey, Crowley, Dirks, and McMahon; the DoD was represented by Fubini, McMillan, Gen. Stewart, Col. David Carter, and Maj. Larry Skantze. Itek had senior representation: President Richard Lindsey, Walter Levison, Richard Philbrick, Edward Campbell, John Wolfe, Frank Madden, and Cal Morser and his project staff. After an extensive briefing and tour, with many questions raised by Fubini, DCI McCone asked Itek some searching questions of his own. Among these, he asked if this system was the very best the company could do! Lindsey replied that it was "not a perfect system but another year of study would only produce marginal gains." McCone then asked, "Is this the best approach?" Lindsey replied, "Yes, considering the constraints." Wolfe said "Yes, at the moment." Levison said "Yes, within a given set of constraints, this comes close to optimum, considering technical reach, manufacturability, and operability." The record of the meeting notes that Itek "had been forced to say that the technical approach was the right magnitude and the approach was optimum."[98]

While Land was willing to provide his own technical counsel to the CIA, he felt that the panel should also be exposed to the other search system studies (S-2) by Eastman Kodak, Itek, and Fairchild Camera and Instrument Company) "in order to make a balanced evaluation."[99] Knowledge of S-2 efforts was seen as relevant to the Land panel, especially since McMillan had outlined them to McCone in a 22 January 1965 letter (which McMillan had intended to serve as background prior to a detailed briefing to McCone and Vance scheduled for 2 February 1965).[100] To complete the information exchange, on 11 February 1965, Wheelon forwarded to the DNRO work statements of the CIA's FULCRUM study contracts for the camera (at Itek and Perkin-Elmer), alternative fast film transports (STL and RCA), systems engineering and assembly (SEAC)[101] (STL), spacecraft (GE), and recovery vehicle Avco).[102]

Land, whose panel was scheduled to meet in the Boston area on 23 and 24 February, asked that terms of reference for his panel's deliberations be established clearly. McCone, who had expressed the view that the panel would be acting as a technical advisor to Vance and himself (the NRO ExCom), agreed to go to Boston on 23 February to clarify the terms of reference and to summarize USIB requirements for a new search system. Carter invited Vance to join McCone.[103]

McCone and Wheelon had done their FULCRUM homework and, barring unplanned developments, were confident of favorable consideration by the panel.

The Itek Episode

In late 1964 and early 1965, CIA-Itek relations were not at their best. The CIA was (naturally) pushing hard to ensure that its contractors' work supported a go-ahead decision on FULCRUM. In September 1964, Itek had proposed a work statement that would encompass work both on the earlier-selected twin 60-inch f/3.0 optic camera design and on a 1/4 focal-length system (later changed to a 1/3 system).[104] After Itek demonstrated adequate availability of personnel, CIA had authorized work on both configurations "emphasizing the twin optical bar has priority."[105] Subsequently, in November, when Itek fell behind schedule, the CIA, after consideration of the pros and cons involved, cancelled the 1/3 focal-length effort. Itek considered this a serious mistake and protested the cancellation, to no avail. Relationship problems between the CIA and Itek continued, and on 11 January 1965, a discussion was held among CIA FULCRUM personnel and Walter Levison and Richard Philbrick concerning prerogatives of the program.[106] At a subsequent informal meeting on 16 January 1965 at the residence of Itek's president, Frank Lindsey, there was "every indication that earlier differences had been resolved and management was most anxious to get on with" the job.[107] One technical issue which persisted concerned the angle through which the camera system would scan. The CIA had wanted—and Itek had proposed—a scan angle of 120 degrees (60 degrees each side of nadir). Subsequently, Itek became concerned that this angle was too large and seriously prejudiced the FULCRUM design; on 19 February 1965, Lindsey sent a letter to Wheelon to this effect.[108]

The 120-degree scan issue was a critical element at a weekend meeting in Washington on 21 February, which Wheelon had called to review the planned Itek presentation for the all-important Land Panel meeting on the 23rd. At issue was whether or not Itek was required, by contract, to design for the 120-degree scan. The Itek representative was program manager John Wolfe. He recalls that Wheelon asked Maxey and Dirks whether they considered the 120-degree scan "a requirement." They replied in the negative. At this juncture John McMahon joined the meeting and was asked the same question by Wheelon. McMahon, who handled contracts and administration, replied that the 120-degree scan was a contractual requirement. In the ensuing discussion, Wolfe was told that this issue was inappropriate for the Land Panel briefing.[109] Wolfe was sufficiently concerned that he contacted his boss, Levison, who was on an unrelated business trip to Chicago with Lindsey and Frank Madden of Itek. The matter was of such importance that Lindsey, Levison, and Madden discussed it for two hours.[110]

On Tuesday, 23 February, the Land Panel convened at Itek's Boston facility for a briefing on FULCRUM work and on the results of the search-system studies (sponsored by the Air Force and done by Kodak, Itek, and Fairchild).[111] The next morning, 24 February, at a breakfast meeting, the Itek managers concluded that circumstances were such that they could not retain their "technical integrity" if they continued (sole source) participation in the FULCRUM program.[112] The painfulness of this decision to a small company was evident, considering that the anticipated FULCRUM effort would be four times the size of Itek's then-current work on CORONA; furthermore, a refusal to continue on FULCRUM would certainly not endear Itek to its best customer—CIA.

A final decision was made that afternoon. It was agreed that Lindsey would inform DCI McCone and that Levison would inform DNRO McMillan and Edwin Land of the Itek decision as soon as possible.[113] Levison called Col. Paul E. Worthman, Chief of Plans on the NRO Staff, and made "the following remarkable announcement: 'For a multitude of reasons, Itek has come to a corporate decision that it cannot accept the follow-on to FULCRUM, even if it is offered.'" The decision was not his, but was that of the company and he stated "that there were *no* conditions which would change this attitude."[114] Levison asked Worthman for advice on how to handle this obviously awkward situation. Worthman urged Itek to advise McCone (or whoever was acting in his stead) with utmost dispatch, particularly as the Land Panel was in the process of issuing highly influential recommendations on the future of FULCRUM. Shortly thereafter, Levison called Worthman again to report that Lindsey, unable to reach McCone, had advised John Bross, a senior member of the DCI staff, of Itek's decision.[115] Levison asked Worthman to arrange a meeting with Land and McMillan; Worthman contacted McMillan and urged him to call Levison.

Consequently, late that afternoon, Levison and Wolfe met with Land and McMillan at the Polaroid Corporation in Cambridge (where the panel was meeting in executive session). When Land and McMillan came out of the conference room to speak with Levison, they were joined by Wheelon, who had been sitting with the panel. (Wheelon's presence caused Levison to approach the discussion more formally than he had intended; Levison, a long-time acquaintance of Land, had hoped to keep his words informal and off the record.) Levison announced the Itek decision;[116] he added that although Lindsey had not been able to reach McCone (reaching Bross instead) Lindsey and Philbrick were on their way to Washington, hoping to see McCone that evening.[117] In a subsequent discussion with McMillan, the Itek representatives said they believed that "they could not maintain 'technical integrity' if they undertook a development project for FULCRUM with as little technical control over the project as they had been allowed during their work up to this time. Itek felt that the rotating optical bar technique to be used in FULCRUM could not be justified unless there was a firm requirement for scan angles of 120 degrees or more."[118] To complicate the matter, DNRO McMillan, in a 25 February 1965 memorandum[119] for Vance, advised him of an earlier meeting with Levison. At that time, McMillan had expected to recommend to Vance and Defense Secretary McNamara the development of a general search camera system *other* than those being studied by Itek (either for CIA or the Air Force). He felt that the Itek staff should be aware of his views so that it might have an opportunity to present Itek's side of the matter.

Whether or not the DNRO's views had an effect on Itek's conclusion to withdraw from FULCRUM cannot be determined. There exist, however, some interesting, but erroneous, views of McMillan's role in the events of late February 1965. CIA records of that period contain the statement "A year later, it was learned by CIA that the day before the briefing of the Land Panel in February 1965, the DNRO (McMillan) had given a development contract to Eastman Kodak for the follow-on search satellite system,"[120] that is, a program go-ahead. The only thing that did happen in the Program A (SAFSP) efforts on a new search system was the May 1965 transfer of the S-2 effort from the applied research/advanced technology category under SAFSP-6 to project status under ▓▓▓▓.[121] McMillan had authorized planning for S-2 as a system, but had limited all work to a study level "pending an official system go-ahead." Clearly, McMillan would need the NRO ExCom's approval for a new system start and, since the DCI was a member of ExCom, it is difficult to understand how the CIA came to believe that McMillan had authorized a system go-ahead without McCone's knowledge. The record indicates that McCone was too deeply involved in NRO matters and too supportive of Wheelon's FULCRUM efforts to be unaware of, or to countenance, an independent step by McMillan. Furthermore, in September 1965, not only was Eastman Kodak not developing the S-2 camera payload, but also its study effort in S-2 had been sharply curtailed and it had been directed "to submit a plan for the early termination of all S-2 activity at Eastman Kodak and continuance of the Eastman Kodak design at Itek."[122] All Kodak work on S-2 ended by early 1966.[123]

Perkin-Elmer Joins the FULCRUM First Team

While the situation was complex (and the reasons for Itek's action equally so), the net effect of these incidents was a slowdown in the pace of FULCRUM. The CIA had hoped and expected that the Land Panel findings would be the basis for early approval of FULCRUM by the ExCom. In order to preserve FULCRUM sensor work and the momentum of the project, the CIA quickly arranged to transfer Itek's work to the Perkin-Elmer Company of Norwalk, Connecticut; Perkin-Elmer had been under CIA contract, as a backup to Itek, since June 1964.[124] It had not been supported at the same level as Itek and, therefore, had to make up for much lost time. John McMahon recalls that when the NRO had previously given him an additional ▓▓▓▓ dollars to augment the FULCRUM effort, he had allocated ▓▓▓▓▓▓▓▓▓▓ to Perkin-Elmer.[125]

The CIA action to strengthen Perkin-Elmer activity was initiated at two high-level management meetings. At the first, Maxey and Dirks met with Robert Sorenson, vice president and general manager of the Electro-Optical Division, and Dr. Kenneth MacLeish, vice president and director of engineering, Electro-Optical Division. Dirks asked if Perkin-Elmer could step up its effort on the FULCRUM program and assign Milt Rosenau as the program manager. Sorenson replied, "Yes and yes—unequivocally." The CIA representatives did not explain why there was a change of direction, only that it was a matter of great urgency.[126] Shortly thereafter Wheelon met with Chester Nimitz, Perkin-Elmer President and Chief Executive Officer, and asked if Perkin-Elmer could take over the program started by Itek. Nimitz agreed to accept the challenge.[127]

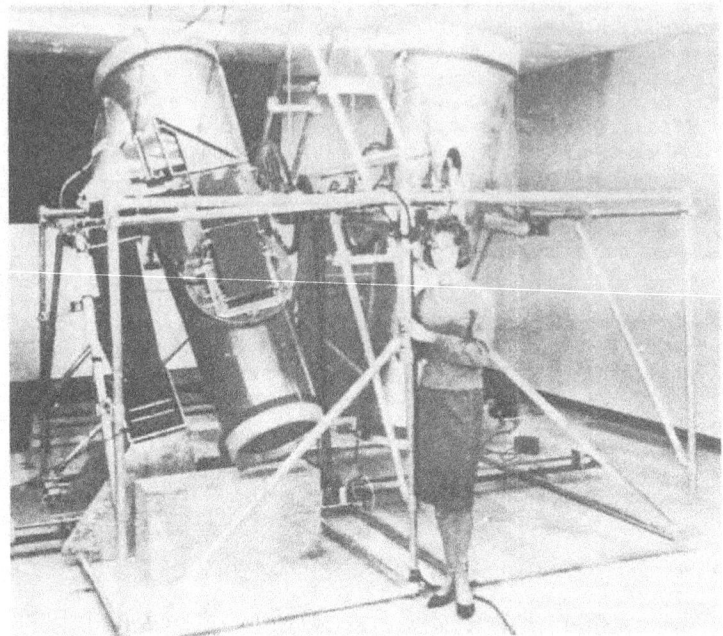

Perkin-Elmer FULCRUM Camera Optics

The Land Panel Recommendations

On 26 April 1965, the Land Panel issued the findings of its 23—24 February deliberation. In evaluating FULCRUM and similar studies at SAFSP, the panel had considered the following questions:

1. How confident can one be that the device shall meet the performance goals?

2. Are there critical technical problems in any one of the proposed systems, the solution of which is not in hand?

3. Is there a likelihood that unforeseen technical problems will be encountered in carrying a particular design to completion and operation?

4. In the light of one's judgment on the preceding questions, how great is the risk of serious delays in reaching operational status and assured operational reliability?[128]

With regard to FULCRUM, the panel concluded that very significant progress had been made on key technical problems which had been identified in the panel's June 1964 evaluation. The mechanical aspects of rapid film transport appeared to be under control and a simpler film path had been conceived. While the dynamic problems of an earlier configuration had been circumvented in the present design, the issue of rotational stability affected by the loading and unloading of very large film spools, particularly with respect to reliable control of the spool's dynamic balance throughout a mission, remained an open concern. A majority of the panel concluded that unless the 120-degree scan angle was an absolute requirement, a revolutionary development was not mandatory and an evolutionary approach at lower risk was preferable. Finally, it concluded that the S-2 systems were reasonably conventional in concept, representing "a relatively short evolutionary approach from present practice."[129]

Land, in an individual statement, held that "although this system (FULCRUM) may not be optimum, the good progress to date and the more thorough system analysis which has been done in this system, compared with others, justify at least tentative authorization for full-scale development. It should be remembered that any of these systems, at anywhere near the claimed cost, will actually save money over the present operations, in addition to contributing greatly to the national security."[130] The full panel report concluded that: "The investigation undertaken in FULCRUM was valuable, informative, and stimulating, even though it does not seem prudent to push FULCRUM as a whole to conclusion. Far from regarding FULCRUM as something that should not have been undertaken, we feel it is exactly the kind of investigation that will be repeatedly needed and that its scope is probably the necessary one for evaluation of any worthwhile fresh approach."[131] It was clear that the panel's report, despite Land's position, did not support early approval of a development go-ahead for FULCRUM.

Shortly after the issuance of the Land report, McCone resigned as DCI, returning to the industrial sector; he was replaced by Vice Adm. William F. Raborn, (USN-Ret), who had managed the Navy's highly successful Fleet Ballistic Missile Program. McCone's deputy, General Carter, became director of the NSA and was replaced as DDCI by Richard M. Helms. Because of his fresh viewpoint and long background as a military officer, Raborn tended to work more harmoniously with the DoD elements of the NRO than did McCone—who as DCI strongly supported the CIA in NRO matters.

Fierce Competition on an Uneven Playing Field

The lukewarm Land Panel Report and McCone's departure did not make Wheelon's role any easier. But despite these losses, he continued to press the CIA's case for an enhanced role in satellite reconnaissance research and development. In a memorandum to DDCI Carter, dated 26 February 1965, he requested organizational authority and personnel allotment for establishing a full-scale satellite devel-

DCI William F.
RABORN

opment office within the DS&T.[132] He reminded the DDCI that he had been operating with only a few authorized people, borrowing others where possible; he cited the various tasks facing his small group; he pointed out that in the previous summer (1964) the DCI had told the President's Foreign Intelligence Advisory Board (PFIAB) that he was creating a group within the Agency to handle satellite programs—a group that might grow, eventually, to 20 or 30 persons.

Wheelon's sense of timing was as, always, exquisite. The steadily growing hostility and conflict between the NRO's Programs A and B, the constant battling between McCone-Wheelon and DNRO McMillan, the endless escalation of all basic decisionmaking to Vance and McNamara—these were elements that made bureaucratic warfare a daily circumstance for the NRO. The contrasting serenity and rapport which had characterized the Bissell-Charyk era had not merely evaporated—it was scarcely remembered.

It was not that Bissell had lacked ambition: he reveled in his warm, continuous intimacy with presidential science advisers, top civilian and military leaders at the Pentagon, and scientific savants, such as Land and Purcell. But he had decided, early in his CIA career, that federal position was enhanced and empowered by continuous diligence in avoiding the burden of administering a large organization. For example, Bissell would never have accepted the notion that he needed an in-house capability for developing aircraft and satellites; he was absolutely convinced that he could build these better, faster, and more economically by having other agencies work for him.

With regard to CORONA "ownership," Bissell had expressed himself typically to Air Force Maj. Gen. Jacob Smart, saying that it was his hope "that the CIA's role in this particular activity and others of a similar nature could be progressively reduced and eventually limited to receipt of the operational product, as one of the customers."[133] In summary, Bissell viewed CORONA through the eyes of an experienced intelligence professional: CORONA research and development was simply a wearying, complicated nuisance which had to be tolerated and patiently endured for one purpose only: to acquire photographs of denied areas.

In contrast, Bissell's ("in-part") successors, Scoville and Wheelon, looked on satellite reconnaissance systems with the eyes and enthusiasms of professional engineers who could scarcely conceal their desire to "get into the business." Of the two, only Wheelon had the energy, imagination, and sheer bravado to demand a huge piece of technological action; it was a brilliant set of Wheelon maneuvers that established the CIA as the nation's newest Space Claimant.

Again, Wheelon's timing could not have been better. The NRO administration was becoming increasingly and unnecessarily vulnerable within its own OSD household. Much of this situation was fallout from a change of directors. Charyk had come to the NRO leadership post richly endowed with previous experience in two high-level Air Force positions; he knew how to "work" the Pentagon and Washington scene. He knew, for example, that one does not burden bosses with problems; at most, he might *mention* an issue—particularly if he suspected it could rise to the Secretary of Defense level from outside sources—but he would accompany the hint with assurances that he could and would handle the matter. He invariably carried out those promises, working quietly, deftly, and behind the scenes to achieve his purpose.

DNRO McMillan came to the OSD "cold." His Bell Laboratories experience gave him very little preparation for the Washington arena; he regularly found himself in awkward, lonely situations; he often carried problems, rather than solutions, to the Secretary of Defense; essentially he sought higher-level resolutions to problems he could not solve.

DCI Dulles would never have tolerated space system research and development as a CIA functional goal; it had been his opinion that even the limited participation Bissell provided to the U-2 and CORONA programs was, in the long run, not in the best interests of an intelligence organization. Dulles, in particular, did not like the high visibility which programs like CORONA and the U-2 gave to the CIA budget process. DCI McCone's preferences, however, based on broad experience in industrial and governmental circles, were quite the opposite. One of his most telling strategies was to humiliate McMillian by refusing to discuss reconnaissance satellite matters with anyone except Vance or McNamara (usually the former) and placing his argument in contexts which explicitly discredited the DNRO. It was the sort of uneven situation in which McCone traditionally gloried. Even Presidential Science Adviser George Kistiakowsky had experienced it in his work with McCone when McCone was AEC chairman and had summed up his encounters with the observation, "I wonder when the next knife will be stuck."[134]

The Vietnam War was an additional factor which, indirectly, overshadowed the NRO. McNamara was personally absorbed, full-time, in demonstrating his belief that warfare could be fine-tuned in scope and violence and, indeed, "run" on a day-to-day basis directly from the Pentagon. The extent of his involvement in the war was close to total; one could observe bombing target selections being made daily on the third floor of the Pentagon.

Absent the availability of strong support from his immediate supervisor, to whom could a newly-appointed DNRO turn for strength and counsel? McMillan did not have the advantage of Charyk's carefully nurtured links to the State Department and to the White House. The Secretary of the Air Force—McMillan's "public" supervisor—would be naturally reluctant to share the DNRO's problems, because the NRO belonged, in actuality, to the Secretary of Defense. And the (military) Air Staff, still smarting from the loss of the satellite reconnaissance function, would be hoping, if anything, that the DNRO *would* falter and the organization itself collapse, perhaps reverting to whence it came. As for the Defense Intelligence Agency (DIA)—the DoD counterpart to the CIA—that organization had just been created, amid intense gunfire from the Army, Navy, and Air Force, and would need more time and experience to become much of a voice within the "Community." In fact, in the entire DoD, there was only one person who showed daily interest in a troubled NRO, and that was Dr. Eugene Fubini, the DDR&E—a generalist in hyperenthusiasm—whose "help" frequently augmented, rather than solved, NRO problems.

Understandably, but paradoxically, as good overflight photo-product began to arrive in Washington on a fairly regular basis, many high-level officials no longer felt a need to extend personal support to the program. To the "customer," whether a President or a lieutenant, progress in overhead reconnaissance was reflected in "take," and, from that point of view, the NRO appeared to be doing well and would be assumed capable of proceeding (organizationally) on its own.

Gradually, but inexorably, McMillan sensed that he was standing alone. This feeling was further aggravated by the location of SAFSP, his engineering base, so inconveniently distant in Los Angeles. Originally, SAFSP had been sited in Los Angeles in order to be near the SSD, the aerospace contractors, and, especially—or reasons political and practical—the Aerospace Corporation (the Air Force's successor to the STL as a "captive" engineering organization). Only later did it become clear that, from the point of view of "protecting the franchise" and building strong "futures," SAFSP should have been placed in Washington as close to the DNRO as possible.

The DNRO and his staff were further isolated and handicapped by their own security system. The idea of hiding the NRO within the halls of the Air Force resulted in confusion for both foe and friend. McCone's constant, deliberate usage of "Air Force" as replacement for "NRO" was clever semantic denigration and soon became commonplace in the CIA. The Agency's James Cunningham spoke of the power of names in a staff study in which he ruminated on positive means for improving CIA-NRO relationships; he proposed, as a key recommendation, the desirability of locating the NRO outside the Pentagon's Air Force area *in a building of its own.*[135] The DNRO would have been in a much stronger political position had he sought even temporary shelter with some other organization—like the National Security Agency—rather than "hiding" behind an "Under Secretary of the Air Force" door-plate.

During these turbulent formative years, the NRO Staff suffered chronically from a conviction that eventually a new DNRO, or the Secretary of Defense, or the DIA, or bright young staffers from the Bureau of Budget (or Congress), or any combination of the above, would have enough insight and "clout" to protest CIA ambitions to duplicate existing satellite research and development efforts and to cry "Halt!" In this regard, the staff was not only overly optimistic, it was also badly informed. It did not know, for example, of the long-time mandate to the CIA (from Presidential Science Adviser and PFIAB Chairman Killian and the Land Panel) to set up a strong in-house scientific and technological capability—a mandate which Dulles and Bissell had discreetly ignored, but which was now being accepted enthusiastically by McCone Wheelon. Finally, the DNRO and the NRO Staff placed too much credence in the power of the NRO charter; they revered it and believed that a simple re-write would clarify matters and eventually allot the total reconnaissance satellite franchise to Program A.

Curiously, the strong concerns of the DNRO and his staff were of very limited interest to (SAFSP) Program A, which referred to them, naively, as "political matters," not understanding that Program A itself existed as a "political matter" in a political world. The Program A organization believed it should concern itself solely with operations; its logo could well have been two stars in an Operations Center, watching for the first sign of a newly launched bird on "rev one." Indeed, Program A saw its sole role to be operational "birding;" politics was the business of its Washington "branch." Operating CORONA and GAMBIT was exciting and absorbing; such work was "the now;" devotion to "the now" contrasted with Program A's very casual attitude toward "futures." Even some years later, a Program A Director exhorted his group to bear in mind that "Our job is operations—not advocating new systems." Accordingly, the Program A technical planning staff was abnormally small and under very little pressure to deliver follow-on proposals. In general, Program A's approach to "futures" followed the conservative path of step-by-step improvement of currently operating systems, which essentially meant improving CORONA and GAMBIT. By contrast, the CIA's Program B was inclined toward radical or revolutionary change—maginative creations which intrigued Land, whose instinctive bias in favor of innovation was reflected in the patents he held, and who had a very strong voice in endorsing "futures."

Wheelon understood the overhead reconnaissance "territory" in detail and entirety, drawing upon his extensive experience with intelligence requirements, operations, interpretation, and application. He reviewed existing NRO "franchises," searching for reasonable entry points for an enlarged Program B, some route which would enable his program to compete with, and expeditiously surpass, Program A in operational sophistication. He sensed a rare opportunity provided by the NRO's weakness on "futures" and decided that his main chance lay in engineering radical payloads which would make existing Program A equipment obsolete. One such payload might achieve, *simultaneously*, an improved search and surveillance capability. If one could build that "dream" payload, booster, and spacecraft, "ownership" might come later.

The DNRO and his Staff underestimated the Program B "threat" to the existing NRO. There was no effort to predict or "war-game" Wheelon's possible courses of action, to link the DNRO (privately) to the Land Panel, to counter with a super-panel of one's own, to woo the President's science adviser, or, at the very least, to develop an entente with Wheelon. The outcome of this negligence was ironic: unable to control Wheelon's far-reaching "studies," the DNRO soon found himself actually funding them from the NRO budget—in effect, subsidizing work which would eventually move the CIA from "Space Claimant" to "Space Inheritor."

In-house, Wheelon used a scholarly draft "think piece" to justify an expanded CIA role in satellite reconnaissance. He outlined the history of the NRO and, based upon the extant situation, described various approaches to governmental management of the program, easily making a convincing case for a major CIA role. (His paper contained some convenient errors of fact, for example, crediting the CIA with developing and procuring the spacecraft for CORONA.) His concluding paragraph staked his claim: "All things considered, it is the issue of satellite reconnaissance that has been central to the NRO problem thus far. Only a small portion of this activity—the payload—is at stake, although it is a large stake because it represents the total intelligence consideration. Several solutions are possible. It is hoped that the CIA proposal of orderly development and procurement assignment provides the most flexible solution for a rapidly changing field. If this is not practical, the assignment of all reconnaissance payloads to CIA is the only way to preserve a balance in this situation and ensure a continuing dedication of these satellite collection systems to national intelligence needs."[136]

In July 1965, McMillan sent a status report to Vance and Raborn "on NRO activities toward meeting satellite search and surveillance requirements in the 1967 and subsequent time period."[137] He expressed the view that in-house NRO studies and analyses, coupled with technology development and parametric analysis by the competing contractors, "had progressed to a point that permitted decisions to be made with high confidence about the overall system configuration" and "that the NRO was now in a position to proceed with an orderly program toward a first launch of a new system in the last quarter of FY67."[138]

The timing of McMillan's "wrap-up" action on the new search and surveillance system is interesting; on 10 July 1965 (three days before McMillan's report), President Lyndon B. Johnson announced that he had accepted McMillan's resignation, effective 30 September 1965. During the interim, Dr. Alexander H. Flax, Assistant Secretary of the Air Force for Research and Development, would serve as Acting DNRO.

Alexander H.
FLAX

Section 8

A New Space Inheritor: The CIA

A New DoD-CIA Agreement on the NRO

On 19 July 1965, not long after he became the DCI, Adm. Raborn proposed a new agreement between the DoD and CIA "to govern our relations with the NRO," forwarding a draft for Vance's consideration. He outlined basic principles to be applied to the agreement:

1. The necessity for the existence of an Executive Committee[139] consisting of the Deputy Secretary of Defense and the DCI to provide policy guidance and supervision and to allocate responsibilities under the program as a whole. (Adopting a recommendation by Mr. McNamara, Admiral Raborn proposed that the President's Special Assistant for Science and Technology join such a committee when research and development matters were discussed.)

2. The DCI, in order to be responsive to USIB requirements, should maintain the responsibility of providing specific program guidance to ensure optimum exploitation of satellite reconnaissance missions for intelligence purposes. (Admiral Raborn suggested that the function and basic personnel incorporated in the NRO Satellite Operations Center be returned to and renamed the Satellite Reconnaissance Programming Office.)

3. The potentialities of all agencies of the Government for the design and invention of new concepts and techniques for the acquisition of intelligence through overhead reconnaissance should be encouraged and exploited to the maximum.

4. The engineering development, testing, and production of new systems is normally the responsibility of contracting firms responsible for the design of these systems. Supervision of these contractors should logically be undertaken by the agency with the best facilities and established competence and experience in dealing with these contractors.

5. To a large extent, programs of the NRP are financed with confidential funds expended under the authority of the DCI and Public Law 110. Suitable provision should be made to safeguard the DCI's obligation for ensuring appropriate control and accounting for such funds."[140]

On 13 August 1965, Vance and Raborn signed an "Agreement for Reorganization of the National Reconnaissance Program."[141] It incorporated Raborn's principles in the main, but did make some basic changes which had serious consequences for the CIA's hoped-for role as *system manager* of the new search system. The last brief paragraph of the agreement stated:

F. Initial Allocation of Program Responsibilities

1. Responsibility for existing programs of the NRP shall be allocated as indicated in Annex A attached hereto."[142]

Annex A is interesting in that it deals only with "assignments for the development of new optical-sensor subsystems"[143] and, relative to the new search system, states, in part, that "following the selection of a concept, and a contractor, for full-scale development . . . the CIA will develop the optical-sensor subsystem for that system."[144]

Allocation of responsibility for the remainder of the system is dealt with in subparagraph 1d under paragraph D of the Agreement which concerns, interestingly enough, the role of the NRP Executive Committee. Specifically, it states that "The engineering development of all other subsystems, including spacecraft, reentry vehicles, boosters and booster interface, shall in general be assigned to an Air Force component, recognizing, however, that sensors, spacecraft, and reentry vehicles are integral components of a system, the development of which must proceed on a fully coordinated basis, with a view to ensuring optimum system development in support of intelligence requirements for overhead reconnaissance."[145]

Both "sides" hoped that this carefully-crafted agreement would provide the incoming DNRO with leverage to resolve the bitter, divisive debate between the NRO and the CIA over roles and responsibilities for the new general search system.

The FULCRUM system concept had not received an essential clear-cut endorsement from the Reconnaissance Panel of the President's Science Advisory Committee.[146] The Panel's 30 July 1965 report "reviewed the Perkin-Elmer, Eastman Kodak, and Itek work on high-resolution search systems" and unanimously concluded as follows:

- There is no technical basis for selecting for development at this time one system over any other, nor does the Panel see an urgency for making a selection now rather than, say, six months from now.

- Each system has intrinsic merits which are attractive, but, at the same time, each exhibits certain problem areas of concern to the Panel.

- The efforts of all three contractors should be continued in order to better define the advantages and disadvantages of each system.

The Panel strongly recommended that "all three contractors be funded for an additional three months and that their efforts be focused on further definition of the unique and special features of systems design and on such analyses, tests, and demonstrations which would further substantiate performance claims."[147] It seemed that impetus toward development of a new search system had fallen off; however, the new DNRO ultimately pressed the subject to a conclusion.

At the time of his 30 September 1965 departure, McMillan furnished a report to Defense Secretary McNamara on the status of the NRO and NRP. His comments on the 13 August 1965 Agreement for reorganization of the NRO are of interest. McMillan stated that the new Agreement went less far in actually defining the structure of the NRO than the old 13 March 1963 Agreement. He considered the Agreement less explicit in stating the authorities of the DNRO and too circumscribed in those it did define. While the new Agreement had evidently been intended to palliate some old frictions, McMillan believed it had weakened the NRO considerably, introducing sources of additional friction. He described three specific weaknesses:

- The Agreement was ambiguous in defining the authority of the ExCom.

- It almost completely omitted reference to responsibilities of the DNRO in connection with reconnaissance operations.

- It imposed no obligation upon the CIA, or anyone other than the Secretary of Defense, to provide a focus of responsibility for actions undertaken in the NRP.

In general, McMillan considered the Agreement to have a "trucial character;" it scarcely touched on the substance of the NRP, but rather set up procedures for negotiating the kinds of dispute that had marked the recent past. Its emphasis upon procedure, its severe dichotomy between the CIA and DoD, its detailed specifics for allocating responsibilities for research and development, and its failure to provide any basis for an operating organization simply opened the way to further extensive negotiation on all the important substantive problems still facing the DNRO.[148] (Although the Agreement might well have contained the weaknesses cited by McMillan, it is a fact that, subsequent to its acceptance, working relations between the DoD and CIA appeared to improve.)

On the same day Flax became DNRO, he received a letter from DDCI Richard Helms who "reported the consolidation of all CIA elements supporting the NRO into an organization headed by a Director of CIA Reconnaissance, Huntington Sheldon, who would report to DDS&T Wheelon."[149] (Sheldon was a senior, experienced, and very capable career Agency employee who had the the full confidence of the DCI and DDCI.) The letter also stated that "all CIA satellite activities . . . would be placed in an Office of Special Projects under Mr. John Crowley." Crowley had replaced Jack Maxey who "felt that CIA's role in the satellite program had been so circumscribed by the terms of the agreement that he could not continue to work within such constraints."[150] On 6 October, Flax responded in a positive manner to Helm's letter. Clearly, the new Agreement would improve the operation of the NRP if the individuals involved were so motivated. John Crowley was intent on developing a cooperative relationship between CIA and the DoD.[151] This objective was further enhanced by the fact that Crowley and Flax got along well.[152]

The Technical Task Group and the Project Management Task Group

The first NRP Executive Committee meeting under the new Agreement was held on 6 October 1965. The ExCom was given a brief review of the three cameras under study for the improved satellite photographic system, together with their contract status. Flax described his plan to establish a Technical Task Group, to be composed of representatives of the CIA and SAFSP, under chairmanship of the NRO, which was to prepare a statement of system operational requirements, to recommend the selection of a system configuration, to formulate plans for contractor selection, and to recommend a program plan (including schedule). Flax indicated that he also planned to establish another task group to define project management structure. The ExCom concurred in these actions.[153]

On 15 October, a Project Management Task Group, chaired by Brig. Gen. James T. Stewart (now Director, NRO Staff) and consisting of John McMahon, CIA, and Col. Paul Heran, SAFSP, was established by DNRO Action Memorandum No. 1 for "the development of a project management plan, assigning responsibilities and authorities and defining management channels for the new photographic search and surveillance system."[154] This task group was to recommend alternative project management arrangements and, subsequent to DNRO guidance on results of the first task, prepare a suitable final project management directive.[155]

A second, related group, the Technical Task Group, was set up by DNRO Action Memorandum No. 2, also dated 15 October, which "directed the conduct of those reviews and evaluations essential to a decision to proceed with the development of a new photographic satellite search and surveillance system."[156] This group was chaired by Col. David L. Carter of the NRO staff with Leslie Dirks, CIA, and Navy Capt. Frank Gorman, SAFSP, as members. The charter was explicit:

- Based upon applicable USIB requirements, prepare a statement of system operational requirements for a new satellite photographic search and surveillance system and define the essential technical and operational criteria which must be met by the system.

- Recommend a basic system configuration.

- Recommend the criteria to be used for subsystem design and source selection.

- Formulate a preliminary master project plan (including schedules).

- Prepare necessary project directives."[157]

Richard M.
HELMS

USN Capt. Frank B.
GORMAN

The Report of the Project Management Task Group

The Project Management Task Group, in its report to DNRO Flax on 26 October 1965, considered various forms of program management for the acquisition of the general search and surveillance system. "The task group recommended that either a single project director be assigned from either SAFSP or CIA, or that co-project directors be assigned to carry out the responsibilities of their respective agencies."[158] This equivocal approach was sent by Flax to Huntington Sheldon at CIA, Gen. Martin at SAFSP, and Stewart for comment. The Task Group Report, as such, has not survived; however, comments concerning it are sufficiently detailed to allow a good understanding of the issues involved in this important action. Three sets of comments were returned to the DNRO on 4 November 1965.

Key CIA comments:

- There exist two sets of choices:

    a. how to divide responsibilities for development of the payload;

    b. how the Air Force and CIA would collaborate in executing assigned responsibilities for the program.

- If it were decided that a single project director would manage the new project, then a decision would emerge: namely, whether the CIA or the Air Force should have primary responsibility.

- The most important factor to be considered in carrying out programs under the "new" NRP is the desire of both the DoD and CIA to ensure that the full and creative participation of each organization is totally exercised.

- CIA agrees (with the report) that it is undesirable to have the new system managed within the NRO Staff. CIA also concurs with the rejection of an integrated system project director, which narrows the choice to a single system project director or a split responsibility, a la CORONA. CIA believes there was sufficient analogy between CORONA and the new system to suggest that the new system could be managed successfully on a joint basis. Defined roles and responsibilities, which heretofore had been lacking in CORONA, would materially add to success in the new search and surveillance system.

- CIA argues that if a single organization were chosen to have primary responsibility for the overall management of the new system, the case for assigning that responsibility to CIA is compelling. The history of the CIA study program, dating back to February 1964, supports this argument.

CIA added its comments on three specific items concerning the assignment of responsibilities:

a. System engineering and system integration. CIA considers it essential that specific constraints be placed upon the overall system engineers and overall system integrating contractor. CIA feels it important to delimit clearly the degree to which the system engineering and integration activities impinge upon the responsibilities assigned to other government agencies.

b. Recovery vehicle module. In light of its considerable experience with CORONA, CIA feels strongly persuaded to endorse a "unanimous recommendation" that it be responsible for the sensor module which, according to the task group, includes the recovery vehicle module. CIA agrees that—if the recovery vehicles were to be employed in other programs managed primarily by the Air Force—a good case could be made for Air Force procurement on this program.

c. Orbit control module contractor. CIA does not consider it of critical importance to follow the task group recommendation that the orbit control module contractor also build the sensor model structure and perform as system integration contractor. CIA surmises that when the overall hardware flow is examined in detail it might well be more economical and expedient to assign the system integration function to the booster contractor.[159]

At this juncture there is evidence that the Agency, or at least Wheelon, was more concerned over the CIA's overall role in the NRP vis-a-vis the DoD Air Force than over the FULCRUM program, *per se*. This is confirmed in a draft memorandum in which Wheelon responded to Flax regarding the recommendation of the Management Task Group. He stated that "the most important factor to be considered in implementing the new NRO Agreement is the desire of both signatories to insure a creative and full participation of CIA in the NRP as a responsible contributor." He then listed all the "Air Force"- managed programs, large and small, and pointed out that, as of that time, the CIA had responsibility for only part of CORONA and for the new ▓▓▓▓ program. He concluded by saying, "in summary, the most important decision which you [Flax] face is—how to preserve appropriate CIA participation in the satellite portion of the NRP."[160] This view was consistent with pressures placed upon the Agency by Land and other senior advisors for improving its scientific and technological capability in intelligence collection and evaluation.

Key Program A (SAFSP) Comments

In consideration of management arrangements for any NRP project, the overall objective should be the strongest, most effective management structure possible. In light of the national importance of the projects, Gen. Martin did not believe that any avoidable degradation of this objective could be accepted responsibly or that the basis of any assignment could be one of maximum utilization of resources, equitable distribution of projects or tasks, or the preservation of separate organizational identity and/or prerogatives of the participating agencies.

- Overall project responsibility and corresponding authority, including responsibility and authority for overall system engineering and system integration, must be delegated to a single person who is organizationally and geographically located and appropriately chartered with respect to the resources involved, such that he can effectively control those resources, as necessary, to carry out his responsibility.

- No management responsibility or authority should be retained by the parent agency, as such (for example, the Air Force has no management responsibility or authority over NRO projects assigned to SAFSP).

- The person having overall responsibility—and any persons he designates—must have unrestricted access to all contractors and facilities participating in the project and all information concerning all aspects of the project. He must have authority to determine need-to-know—for these personnel—for any information concerning the project and authority to grant any clearances necessary to personnel he determines to meet published clearability requirements.

- For projects where divided management is directed, the person having the overall responsibility must be delegated corresponding authority over all participants in both agencies, established by directives in each agency and sent to all persons concerned.

Martin opined that the range of the task group's excursion into management approaches (some of which were excluded by the NRP Agreement) and inconsistencies between the task group's stated conclusions and supporting rationale were such as to render the task group recommendations, *per se*, of questionable value.[161]

Key NRO Staff Comments

The position of the NRO Staff was that:

- The Agreement reflected an obvious desire to maintain organizational identity and responsibility. The casual discarding of the fully integrated SPO solution was deplorable and distressing. The fully integrated SPO approach to management was the only valid solution for a complex system development; all alternatives proposed were, in effect, committee-management proposals, full of inherent weaknesses.

- There must be a single, authoritative, responsive system project director.

- There should be established a fully integrated SPO (which collocated all CIA-DoD engineering, procurement, and security people in one office, and empowered those people to speak authoritatively for their "sponsors").

- Although, overwhelmingly, the management capability to do the job was already within SAFSP, total system assignment to CIA would be vastly more effective than the "idealistic but impractical social venture" proposed in the task group report.[162]

Key NRO Staff Director Comments

Gen. Stewart found it necessary to offer his own comments:

- While he strongly desired the fully integrated SPO approach, he recommended against its selection, in view of the apparent intent and specifics of the NRP Agreement.

- He recommended selection of the so-called segregated SPO approach, with overall system responsibility and the System Program Director, assigned to SAFSP.

- It was his view that SAFSP was the only logical choice for overall system responsibility and for providing the SPD.

- He had no firm convictions on the matter of collocation; there was no question about the necessity for collocating a "line" Deputy SPD.[163]

- The CIA office of special projects (OSP) should be charged with developing the sensor module. This would enhance the Government's ability to hold the camera contractor responsible for key factors associated with proper camera functioning.

- He recommended against inclusion of the camera subsystem and a combined sensor/RV module in the sensor-source selection.

- It was his recommendation that the OCV contractor also build the sensor module shell and RV module and be the system integrator.

- He believed an early selection of the system engineer (regardless of management approach) was vital to the work of the three source selection task groups.[164]

After all comments had been made, it was clear that the Project Management Task Force had not yet provided the DNRO with a simple, effective management approach to the new system, particularly considering the policies implicit in the August 1965 NRP Agreement.

Despite agreement, within Colonel Carter's Technical Task Force, between CIA and SAFSP on the idea of a sensor module which included the reentry vehicle, the DNRO was not persuaded, and he ultimately rejected the modular approach in favor of an integrated approach.[165] Faced with the lack of consensus on the "right" way to do the project, Flax had to devise his own plan for the management and technical approach. This complicated chore came at a particularly busy time for Flax. Unlike his predecessors, he was not directing the NRO as Under Secretary of the Air Force, but as Assistant Secretary of the Air Force (R&D) and, as such, had his plate more than full of Air Force development programs. One of these demanding a great deal of personal attention was the (then-designated) TFX airplane, a tactical fighter-bomber which Secretary of Defense McNamara had decreed to be common to both the Air Force and Navy. On top of this, President Johnson's concern that the US Supersonic Transport Program be given professional guidance caused him to ask McNamara to have someone keep an eye on the program, and, because of Flax's past experience at Cornell Aeronautical Laboratories, McNamara turned to him to satisfy the President's concern (even though the Supersonic Transport Program was not a DoD or an Air Force project).[166]

Despite such extraordinarily heavy non-NRO demands upon his time, Flax continued to make progress on the new space search and surveillance system. On 1 April 1966, he forwarded to Sheldon, Martin, and Dr. Donald Steininger (of the PSAC staff) copies of a plan for the new system, which he designated HELIX.[167] This plan had a covering memorandum which requested that: "If you are aware of any factors not previously called to my attention which might impact on the attached, please so advise me as soon as possible and I will consider possible adjustments. Otherwise, I anticipate sending this package to the ExCom in the afternoon of April 5th."[168] (The final plan, as forwarded to the ExCom, was assembled by Flax, personally, and coordinated in draft form with Sheldon and Hornig.[169])

Dr. Flax's proposed ExCom submission reviewed the activity of the NRO staff, the CIA, and SAFSP in carefully evaluating all aspects of the proposed new system. Specifically, it discussed one of the more difficult problems — to devise a technique which would permit the equitable competition of three[170] proposed cameras (designed against varied technical and operational requirements), all of which were at different stages of analysis, creation, and demonstration. He also described the general system configuration to which NRP participants had agreed and which he was recommending for adoption.

Flax recommended a management approach that would make the CIA OSP responsible for the entire sensor subsystem and SAFSP responsible for the remaining system elements. The Director, SAFSP, would be designated system project director, (SPD) responsible for overall system engineering, system integration, and integrated project management. Flax concluded that this assignment of responsibilities–generally in accordance with assignments described in the August 1965 NRP Agreement–would provide effective system management."[171]

In responding to Flax's inquiry, Gen. Martin held that it was important to collocate the program management team "regardless of the specific assignment of responsibilities in the split-management structure."[172] He felt that "regardless of other details of the split-management structure, liaison officers are highly undesirable at any location; they will impede rather than help achieve the rapport essential to a successful development." He was also concerned that "the schedule contemplated in the package leaves no alternative but to employ letter contracts" (as opposed to negotiated definitive contracts). He pointed out that, although contract definitization after source selection would add six months to the schedule, "since the stated requirement is no more urgent now than it was a year ago, and in view of the

USAF Maj. Gen. John L. MARTIN, JR.

non-technical delay already accepted during this past year, it is not obvious to us here that this relatively small additional delay would be unacceptable." He was concerned that the proposed role of the SPD in carrying out his overall system engineering/technical direction responsibilities had "restrictions which seem most unrealistic and unnecessary." He then presented reasons for suspending "the restrictive language concerning the authority of the SPD during an operational mission."

In CIA response, Sheldon held that the proposed scheme of management and organizational responsibilities for HELIX "raises a problem of such magnitude that it must be resolved before other aspects of the program can be meaningfully reviewed" and specifically cited concern "over the problem of interface between the responsibilities assigned to SAFSP (Air Force) and CIA."[173] Sheldon took direct issue with the DNRO when he told Flax ". . . with CIA's in-house technical personnel and its relationships with contractors built up over the years, it possesses a capability of program management commensurate with that of SAFSP . . . . I cannot accept your statement that SAFSP is the only NRP component of the NRO possessing the personnel, facilities, operational resources, experience, and technical competence to be designated SPD for the new general search and surveillance satellite system."[174]

On 22 April 1966, the DNRO submitted, for ExCom consideration and approval, his complete proposal for the new general search and surveillance system (still under the HELIX caveat). The package included:

- A System Operational Requirement (SOR) document which established technical and operational criteria for the entire system.

- The sensor subsystem Request for Proposal (RFP) to be issued to Itek and Perkin-Elmer. (Flax had eliminated Kodak earlier. Kodak was already assigned the

- A management plan for the development, production, and operation of the new system. This included the assignment of responsibility to the CIA for the sensor subsystem and to SAFSP, as SPD, for overall system engineering and system integration, for the satellite basic assembly, the reentry vehicles, and the mapping camera.

- A group of five papers establishing the rationale for key portions of the SOR, RFP, and management plan and explaining requirements, system life considerations, recovery vehicle considerations, measurement of system effectiveness, and system management.

- A schedule of near-term planned NRO actions.

Dr. Flax specifically requested ExCom approval of the system concept, the management plan, and the fundamental principles set forth in the SOR and RFP.[175]

The day before the ExCom meeting, Flax addressed a separate memo to Deputy Defense Secretary Vance, advising him of the reactions he had received to his proposed HELIX plan and of issues likely to be raised at the meeting.

ExCom Approves the HEXAGON Management Plan

At its 26 April 1966 meeting, HELIX/HEXAGON[176] was the first item on the ExCom agenda. "Vance proposed at the outset that, after such discussion as was necessary, Adm. Raborn, Dr. Hornig, and he meet in executive session to make the required decisions. Adm. Raborn and Dr. Hornig agreed."[177] Following the HEXAGON briefing, Adm. Raborn said that he had only one major recommendation to make on the proposed management plan: that the sensor subsystem definition be modified to assign CIA responsibility for the structure which enclosed the sensor subsystem, as well as responsibility for the development, production, and integration of the stellar-index (mapping) camera.[178]

Before the end of the HEXAGON discussion, Sheldon, who was still in attendance, suggested a need for further examination of difficulties which the CIA believed might result from the plan: would both the SAFSP and CIA project offices be authorized to grant HEXAGON clearances? Would each honor need-to-know determinations on the part of the other? ExCom asked Sheldon and Flax to examine this matter.[179] Following the closed executive session, Vance advised Flax "that the Executive Committee had approved his HEXAGON program proposal as submitted (without the Raborn-recommended change)."[180]

Thus, finally, more than two years after the original FULCRUM planning, formal authority was granted to proceed with developing a new search and surveillance satellite system. The CIA's role had been reduced from total system development to responsibility for the main photographic sensor. The compromise on management structure made management more complex, perforce, than it would have been under a single organization. It was clear that a great deal of work needed to be done by *both* government managers before the program could become successful. Flax's compromises did not resolve all potential questions on HEXAGON, but they did resolve some earlier conflicts. "Turf battles" had been reduced to the point where the program could proceed.

Section 9

The HEXAGON Development Program

At the same time the DNRO issued his management directive for HEXAGON, he also provided the following "Systems Operational Requirements for the New Search and Surveillance System:"

   a. To provide "an optimum capability for fulfilling the search and surveillance objectives specified for the time-period beginning in 1969 . . . ."

   b. "Systematic search of some 12 million square nm may be required semiannually."

   c. "Periodic surveillance is required of previously known specific objective targets at a ground resolution sufficient to detect and analyze changes in status or capability of a target."

   d. "Numerically, coverage approaching a total of ▮▮▮ specific targets may be required with coverages of various numbers required at two months, quarterly . . . ."

   e. "During periods of crisis . . . coverage of any selected area . . . to prove effective . . . must be flexible, i.e. capable of prolonged standby prior to launching, rapid response after decision to launch . . . . In addition, the overall system must be designed for minimal time between launching, recovery, and delivery of photography to the user."

   f. ". . . . ground resolution from perigee altitude 2.7 ft, or better, at nadir."[181]

These requirements were frequently abstracted as "development of a single capability for search and surveillance with continuous stereoscopic ground coverage equivalent to KH-4 [CORONA] and a resolution equivalent to KH-7 [GAMBIT]."[182]

Under the management directive, the program was divided, with the CIA responsible for developing the Sensor Subsystem and SAFSP responsible for the remainder of the system; that is, the satellite basic assembly (SBA), recovery vehicles (RVs),[183] Stellar Index Camera (SI),[184] and integrating these parts into a complete system. This arrangement proved to be extremely complex. When technical or managerial differences arose between the Sensor Subsystem Program Officer (SSPO), ▮▮▮▮▮▮▮ (CIA) (SPO), and the System Program Officer (SPO), Col. Frank S. Buzard (SAFSP), the only common arbiter was, necessarily, the DNRO. Since both ▮▮▮▮▮ and Buzard were reluctant to refer problems to the DNRO, long and intense negotiations were required to solve problems.

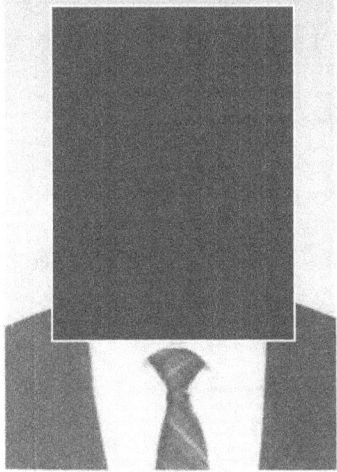 

USAF Col. Frank S. BUZARD

These management arrangements gave the system program director (Director, SAFSP) responsibility for overall system engineering (including master system specifications) and integration, preparation of the system for launching, the actual launching, on-orbit operations, and recovery activities. There were, of course, restraints on the scope of the SPO authority in certain areas. For example, the overall system engineering and integration responsibilities of the SPO would include all interfaces with the sensor subsystem, but not system engineering or technical direction for the sensor subsystem itself. On the other hand, the SPO, in the exercise of interface responsibility, was expected to meet the basic structural, dynamic, and thermal power requirements of the sensor subsystem.[185]

It was stated clearly in the management documents that HEXAGON would be an integrated system in which the sensor subsystem would be embedded within the satellite vehicle, in contrast to being a separate, bolted-on "payload" section. This feature added to the complexity of the source-selection process. The two sensor competitors had generally ill-defined and widely divergent structural, electrical, and thermal interfaces with the satellite. Similarly there were four satellite competitors with widely differing concepts. Both SAFSP and CIA recognized that after the contractors were selected there would be a period of intense interface negotiation, compromise, and modification of design to create an integrated system. They estimated that this negotiation would take about three months.

The HEXAGON Source Selection Gets Underway

In their eagerness to get the system underway, ███████ and Buzard immediately began the source selection process for their parts of the system. This was done in a spirit of cooperation and mutual assistance between the two offices but without a common understanding of system configuration or how hardware would be assembled and tested on its way to the launching site at Vandenberg AFB (VAFB). Representatives of the CIA did participate actively in SAFSP's satellite and RV source selections; similarly SAFSP people worked with the CIA on the sensor subsystem selection.

The schedule for these activities was geared to an October 1966 decision date for contractor awards.

Source Selection Schedule

Part of System	RFP Proposal Issued	Proposal Due	Brief DNRO On Evaluation	Actual Decision Date
Sensor Subsystem	23 May 1966	27 Jul 1966	1 Sep 1966	Oct 1966
Satellite Basic Assembly (SBA)	16 Jun 1966	22 Aug 1966	26 Sep 1966	Jul 1967
Recovery Vehicle (RV)	19 Jul 1966		20 Oct 1966	May 1968
Stellar-Index Camera	24 Aug 1966	17 Oct 1966	4 Nov 1966	May 1968

The SSPO issued RFPs to Perkin-Elmer and Itek for the sensor; potential contractors for the SBA were LMSC, GE, McDonnell, North American, and Hughes (which decided not to participate); for the RV, GE, Avco, McDonnell, Lockheed Missile & Space Company (LMSC; which decided not to participate); for the SI, Itek and Fairchild.

In general, the source selection process proceeded on schedule; however, when briefed on the RV and SI results, the DNRO directed that competitors be allowed to correct their proposals to eliminate weaknesses found by the Source Selection Boards. The revised proposals went through the source selection process again and results were given to the DNRO on 7 March 1967.

Source Selection Candidates and Results

Part of System	Request for Proposal	Proposal Response	Selected
Sensor Subsystem (SS)	Perkin-Elmer	Perkin-Elmer	Perkin-Elmer
	Itek	Itek	
Satellite Basic Assembly (SBA)	LMSC	LMSC	LMSC
	GE	GE	
	North American Aviation	NAA	
	McDonnell	McDonnell	
	Hughes	Declined	
Reentry Vehicle (RV)	LMSC	Declined	
	GE	GE	
	Avco	Avco	
	McDonnell	McDonnell	McDonnell
Stellar-Index Camera (SI)	Itek	Itek	Itek
	Fairchild Camera & Instrument Company	Fairchild	

Perkin-Elmer, proposing a newly organized Optical Technology Division (OTD) under leadership of W. Richard Werner and Michael Maguire, responded to the sensor subsystem RFP with the FULCRUM camera system previously described: two counter-rotating optical bars, an oscillating platen, the film supply oriented in the roll axis, associated electronics, and a film-transport system, with the film to be recovered in two RVs. This entire assembly was to fit into a vehicle with an outer diameter of 100 inches[186] and a length of 170 inches.[187]

LMSC, under the leadership of its program manager, Dr. Stanley I. Weiss, responded to the satellite RFP with a vehicle that was 10 feet in diameter and had an overall length of 46 feet 10 inches, of which a 33-inch section was devoted to the satellite-control unit (containing the equipment rack, the propulsion module, and the reaction-control module), 15 feet to the sensor subsystem, and 25 feet 4 inches to the recovery section if it had four RVs, and 20 feet 5 inches if it had two RVs. The total vehicle weight was approximately 16,000 pounds, including the four RVs and all expendables. Lifting this weight was well within the capabilities of the Titan-IIID booster.

In April 1966, when the DNRO gave program go-ahead, all those involved in the program—the SPO, the SSPO, and the various potential contractors—assumed that actual development of the system would begin about 1 October 1966, when major source-selection activities had been completed. Unfortunately, such was not the case. Sensor subsystem go-ahead was given on 7 October 1966, SBA (the spacecraft) on 20 July 1967, and the recovery system and the stellar terrain camera were not approved until May 1968! The nine-month delay between the start of sensor development and spacecraft development created a number of technical problems which added substantially to the cost and time required to develop the system. The design of the sensor system proceeded for almost 10 months on an assumed interface with the spacecraft and the RVs; this design turned out to be incompatible with the design(s) of the rest of the system at a number of critical points.

**Stanley I.
WEISS**

It took another 10 months—from July 1967 to May 1968—to resolve these differences. This intense effort by the SPO, SSPO, and their contractors resulted in significant changes to the hardware designs of both the satellite and the sensor:

a. The film-supply reels for the sensors were reoriented from the roll axis to the pitch axis to simplify the problem of controlling vehicle attitude while on orbit.

b. The midsection of the satellite vehicle was lengthened by 54 inches to accommodate this change.

c. The sensor envelope (the space assigned to the cameras) was established as 110-inch diameter.

d. Electrical power characteristics were brought to a common standard.

e. Sensor command needs were fitted into the planned command system capabilites.

f. Thermal design concepts were rationalized between the sensor and the satellite.

The original FULCRUM and HEXAGON designs recovered the film in two large RVs. While this approach used the simplest film path and added the least weight, it did limit the operational flexibility of the system and increase the average age of the recovered material. (For a 30-day mission, recovery would be scheduled for days 15 and 30.) Although studies were made using as many as 12 RVs, serious consideration was limited to the four-RV when there was an urgent need for the material (photos) on board–seemed to override the increase in complexity and the slight decrease in reliability occasioned by additional RVs. Although the four-RV configuration would require considerable redesign of the film-path into the RVs, the DNRO approved that configuration in July 1967.

After a period of study and negotiation on the factory-to-launching-pad sequence (as with most other problems), the SPO and SSPO arrived at a mutually satisfactory solution. In this case the midsection, built by LMSC, was shipped by C-5 aircraft to Perkin-Elmer, where the cameras and the 1,576- pound, 208,000-foot film supply were installed and tested. The section was then returned to LMSC at Sunnyvale, where it was mated with the rest of the satellite. In the meantime, the RVs would have their film take-up reels installed and carefully aligned at Sunnyvale. The four RVs would be installed in the forward section, which would be mated with the aft and midsections. Once completed, the assembled vehicle would be tested (operated) in thermal vacuum chambers (simulating the space environment) and then shipped to VAFB in a completely flight-configured condition, pad-ready for integration with the Titan-IIID.

Early HEXAGON Development Activities

With the award of a contract for developing the sensor subsystem, consisting of the twin optical-bar cameras and associated film-supply and film- transport system, a period of intense activity began at Perkin-Elmer. At the time, Perkin-Elmer had a business base of $88 million; the HEXAGON program was estimated at ▓▓▓▓▓. The total Perkin-Elmer employment in the Norwalk, Connecticut, area was 2,800 (1,350 of these were in the Optical Group) of which 150 people were involved with HEXAGON.[188]

Manning the program was Perkin-Elmer's first problem: where would it get the numbers of talented people required? Perkin-Elmer's original proposal contemplated growth from 150 to 600 people within four months and to 700 by the eighth month. Perkin-Elmer intended that "additional manning require ments would be met primarily by transfers from the Electro-Optical Division and by an extensive recruiting program."[189] This growth rate soon proved impossible to achieve, and it was not until 15 months later that 700 people were on board (and productive). The basic contributors to the manning problem were the high cost of living in the area, the relatively low salaries offered by Perkin-Elmer, and, perhaps most importantly, the time required to go through essential security investigations and clearance procedures for each individual. As a result of the latter problem, a large pool of uncleared, nonproductive, costly manpower accumulated at Perkin-Elmer during the first year of the program.[190]

HEXAGON System Concept

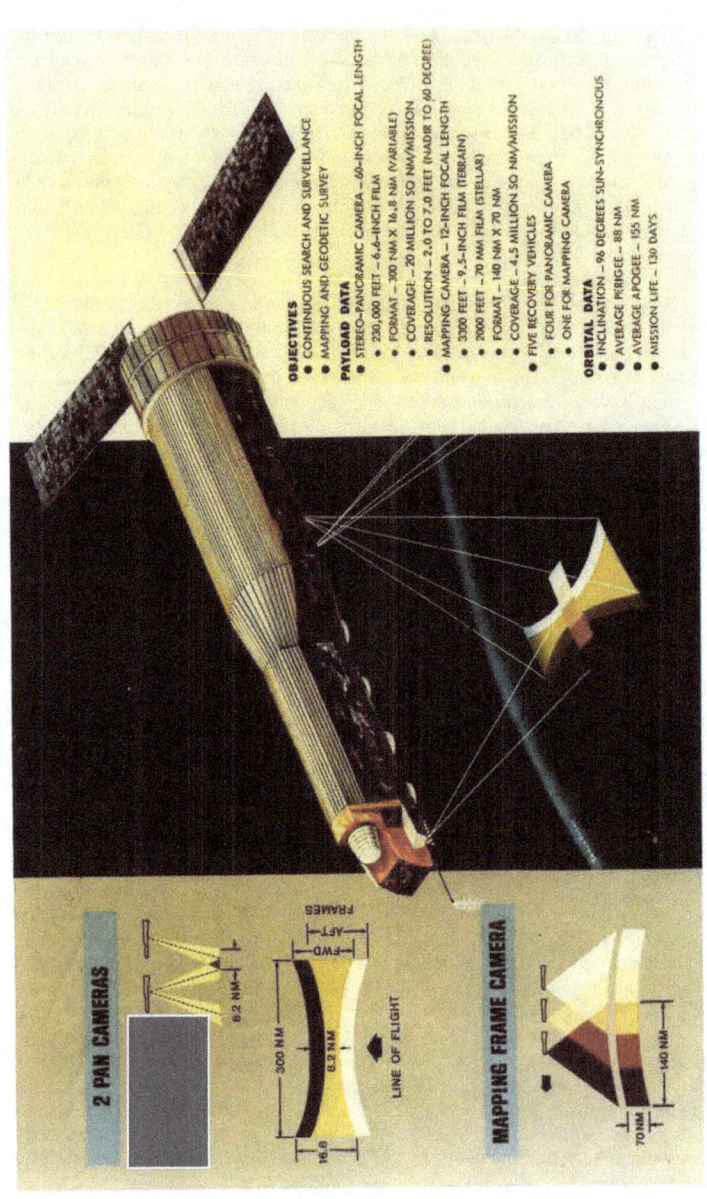

HEXAGON Booster on Launching Pad

Perkin-Elmer's lack of extensive electronic-design experience and its shortage of electronics engineers created very serious problems. When it became apparent that electronic design was falling far behind schedule, Perkin-Elmer subcontracted that work. This caused new problems, since Perkin-Elmer's structure for managing subcontracts was inadequate. Although all of these difficulties were overcome within the first year, they did cause serious slippages in sensor subsystem delivery.[191]

Additionally, the general Perkin-Elmer management structure was inadequate in both scope and experience for a program like HEXAGON. This condition was reflected in the need for two Perkin-Elmer reorganizations during the first year of the program.[192]

Between October 1966, when Perkin-Elmer received contractual go-ahead, and July 1967, when LMSC was awarded the SBA contract, the Sensor Subsystem Program Office of TRW, the systems engineering contractor, played an important, though equivocal, role in the project. Since TRW had the system experience that was lacking at Perkin-Elmer, the SSPO assigned TRW some functions that would normally have been assigned to Perkin-Elmer:

- Definition of the sensor subsystem and its operational requirements

- Preparation of development, acquisition, and operational plans

- Preparation of system specifications

- Definition and control of design interfaces

- Surveillance of the development and acquisition of system elements, including participation in design reviews to verify compliance with system requirements

- Review of equipment integration and acceptance test plans for adequacy, to assure meeting performance specifications.[193]

Perkin-Elmer people quite naturally regarded the (very) active local presence of TRW and TRW's participation in the general design and development process as unnecessary interference; this reaction added a morale burden to an already understaffed and overloaded work force. On the other hand, TRW believed the management at Perkin-Elmer was, and would continue to be, inadequate to the task and that TRW should be given a much stronger role, including technical direction and hands-on management. At one time, TRW suggested to the Sensor Program Office that it become the prime contractor, with Perkin-Elmer as a subcontractor.[194] The SSPO decided to restrain TRW's activity at Perkin-Elmer. When the SBA contractor was announced, TRW became the primary support to the SSPO in the negotiation of the technical interfaces between the sensor subsystem, the satellite basic assembly, and the other parts of the system.

Once the DNRO gave the go-ahead for the satellite contract, the SPO organized a number of interface working groups (IFWGs) to define, negotiate, and police each interface. Each group was headed by the responsible officer from the SPO, with representation from the SSPO, Aerospace, TRW, and each contractor. Initially, these groups—test and assembly; electrical; tracking, telemetry, and control; structural/mechanical and thermal; and operations—met for about one week each month to work on mutual problems. As time went on, the groups resolved many incompatibilities between the sensor and the SBA; however, by April 1968 it was apparent that the process should be ended, since it consumed valuable engineering manpower and delayed overall program progress. In May 1968, the SPO called a "negotiate until complete" meeting in Los Angeles to resolve remaining differences. This meeting lasted four days, ending in agreed-upon documentation which defined the interface between the sensor subsystem and the SBA.[195] In spite of such difficulties, both Perkin-Elmer and LMSC continued developing many critical parts of the system which were not affected by the interface problem.

In July 1968, the DNRO gave the go-ahead for the RV contract with McDonnell (now called McDonnell-Douglas as a result of a merger of the two companies). The interfaces between the RV and the rest of the vehicle proved relatively simple to define.

In the period prior to July 1968, there had been considerable discussion of the need for a stellar-terrain subsystem. In order for imagery to be useful for mensuration purposes (that is, measuring distances and determining the size of objects on the ground) there was a need to record satellite attitude and position information at the instant a picture was taken. In the CORONA system, this had been done by a stellar-index camera—a separate unit which took pictures of both the starfields and the ground, from which vehicle attitude and position could be determined accurately. Film from this unit was then fed into the RVs for recovery along with film from the main cameras. This arrangement also made it possible to prepare maps from CORONA imagery. The mapping community, represented primarily by the Defense Mapping Agency, desired a means of making maps from HEXAGON imagery. It soon became apparent that these two requirements (mensuration and mapping) should be handled separately. The photointerpreters at NPIC needed 3 arc-seconds accuracy for mensuration purposes.[196] Sufficiently accurate attitude determination could be obtained from the attitude-control system telemetry data thereby eliminating (1) the need for the stellar imagery for target location and (2) the complication of recovering this material in the main RVs. Therefore it was concluded that a separate mapping camera would be added to meet mapping requirements.

In late 1967, Deputy Secretary of Defense Paul Nitze requested a study to finalize the nature of the HEXAGON mapping camera, which had been a matter of controversy. As a result of that study, on 12 March 1968 the DNRO directed the HEXAGON SPO to proceed with the development of a system with a 12-inch focal length terrain camera lens (rather than a 3-inch system which had also been proposed). The report noted that the smaller lens system required correlation with panoramic photography to produce medium-scale maps while the 12-inch system would permit production of medium- and large-scale maps without correlated panoramic photography.[197]

Several major system problems still had to be resolved before a final HEXAGON configuration could be established. One of the most difficult of these centered on the recovery of stellar-terrain camera film. Should a portion of this film come back in each of the four main RVs? Should it all come back in the last RV? Should it have its own RV? The mapping camera would have two film supplies: a 9-inch film for the terrain camera and 70-mm film for the stellar camera. To add these complicated, delicate film paths, their take-ups, and their associated electronics to the RV main take-ups would be a formidable undertaking. Choosing to bring all the film back on the last RV would limit the space and weight available to sensor film and create a nonstandard RV. Also, since the mapping camera would probably be on only a limited number of vehicles, the "last RV" option seemed undesirable. Although a separate RV would mean additional weight and an additional recovery operation for the recovery crews, the DNRO eventually approved that solution. This RV module was flown on vehicles 5 through 16. Since film size and weight were compatible with the Mark-V capsule from the CORONA and GAMBIT programs, it was a relatively simple matter to incorporate the unit into the total system. In July 1968, Itek was given the go-ahead for the stellar-terrain camera and GE was directed to modify the Mark-V recovery capsule. The stellar terrain or mapping camera module development was managed by the SAFSP HEXAGON SPO.

As previously noted, while these studies to resolve the SI problem were going on, the DNRO approved award of the RV contract to McDonnell-Douglas and the mapping camera contract to Itek (May 1968). These companies were integrated rapidly into the HEXAGON program, but responsibility for the development of operational software for the system was unresolved and remained a major concern to both the CIA and SAFSP managers.

Evolution of a Design Philosophy

Key members of the HEXAGON SPO—particularly Col. Frank S. Buzard and ▬▬▬▬▬▬▬▬—had had extensive engineering experience in the original CORONA program or in SSD's Agena Program Office. They recalled the many problems of CORONA, a severely weight-constrained, non-redundant system, where the failure of a single component generally led to mission failure. They had seen the extensive redesign of CORONA subsystems: the numerous booster modifications and system improvements that were needed to achieve longer lifetimes on orbit. They knew the hazards—often catastrophic—of making payload or other changes and failing to notify the engineers responsible for system electrical circuitry. With these experiences in mind, Buzard and ▬▬▬ stipulated that "the SV have an 81 percent probability of successful operation for at least 30 days with a goal of 50 days at 80 percent. Furthermore, selections would be based on previously designed and qualified hardware. Redundant wiring would be provided for all critical power and signal leads. And, most importantly, a strong system engineering function would be essential.[198]

The original RFPs and resulting proposals were based on using a Titan-IIID—defined as a Titan-III core with three-segment solids—which would provide a lift capability of approximately 16,500 pounds into the desired orbit. Between the time the RFPs were issued in April 1966 and the go-ahead for the satellite contract in July 1967, it became apparent that this Titan configuration would lead HEXAGON into the same weight-constrained situation that had plagued CORONA. Colonel Buzard recommended to the DNRO that the Titan-IIID be defined as a Titan-III core with five-segment solids. This change, approved by the DNRO on 29 June 1967, increased the lift capability to approximately 20,000 pounds, providing a margin for HEXAGON growth. Additionally, Buzard and ▓ insisted that after system tests had verified compatibility and system integrity, the entire assembled satellite vehicle—SBA, sensor subsystem, and RVs—would be end-to-end tested in simulated mission profiles, including dynamic optical testing in thermal vacuum chambers representing the space environment. During these tests, all the subsystems that could be exercised would be operated to insure a "launch-ready" condition for the satellite delivered to VAFB.

Also, as a result of CORONA and GAMBIT experience, LMSC developed a design philosophy that "no single-point failure shall abort the mission," and "there will be graceful degradation in the event of failure." "No single-point failure" meant, for example, that wires carrying signals from two redundant black boxes had to be in two separate cables with separate routings and grounding points. It meant the creation and detailed review of system wiring and diagrams which would provide end-to-end checks on all electrical power, signals, and telemetry circuits, ensuring that the "no single-point failure" philosophy was carried out in actual design. LMSC also sized many of the critical items—such as fuel tanks—to allow for future growth in orbital life beyond the 50 days specified.[199]

The Factory-to-Pad Process

Perkin-Elmer and the SSPO both wanted to do final performance testing of the sensor subsystem at Perkin-Elmer, after it was installed in the midsection. Once the midsection was mated to the aft and forward section (to form the SV), only minimal camera operation would be permitted. Thus, if a camera malfunction were detected or if any changes were required, the entire midsection would need to be returned to Perkin-Elmer. In contrast, the SPO intended to conduct complete integrated system tests—including acoustic tests to simulate the ascent environment, camera optical performance tests, and on-orbit simulation—prior to shipment to VAFB for launching. This entire testing sequence would require about four months. Thus the SSPO and Perkin-Elmer did not agree with the SPO that there was a need to confirm optical performance of the sensor at LMSC. In SSPO's planning, the final optical testing would be done at Perkin-Elmer, after the sensor had been installed in the mid/section; no real testing would be done at LMSC. In the end, the SSPO and Perkin-Elmer essentially accepted the SPO plan: thorough system-level testing in thermal vacuum chambers, including dynamic optical testing in a special collimator-equipped chamber at LMSC. This capability proved invaluable later in processing the first flight system; when critical camera components failed, they were replaced, and then the integrated system was tested to be certain that camera performance was not impaired.

Titan-IIID Booster Vehicle

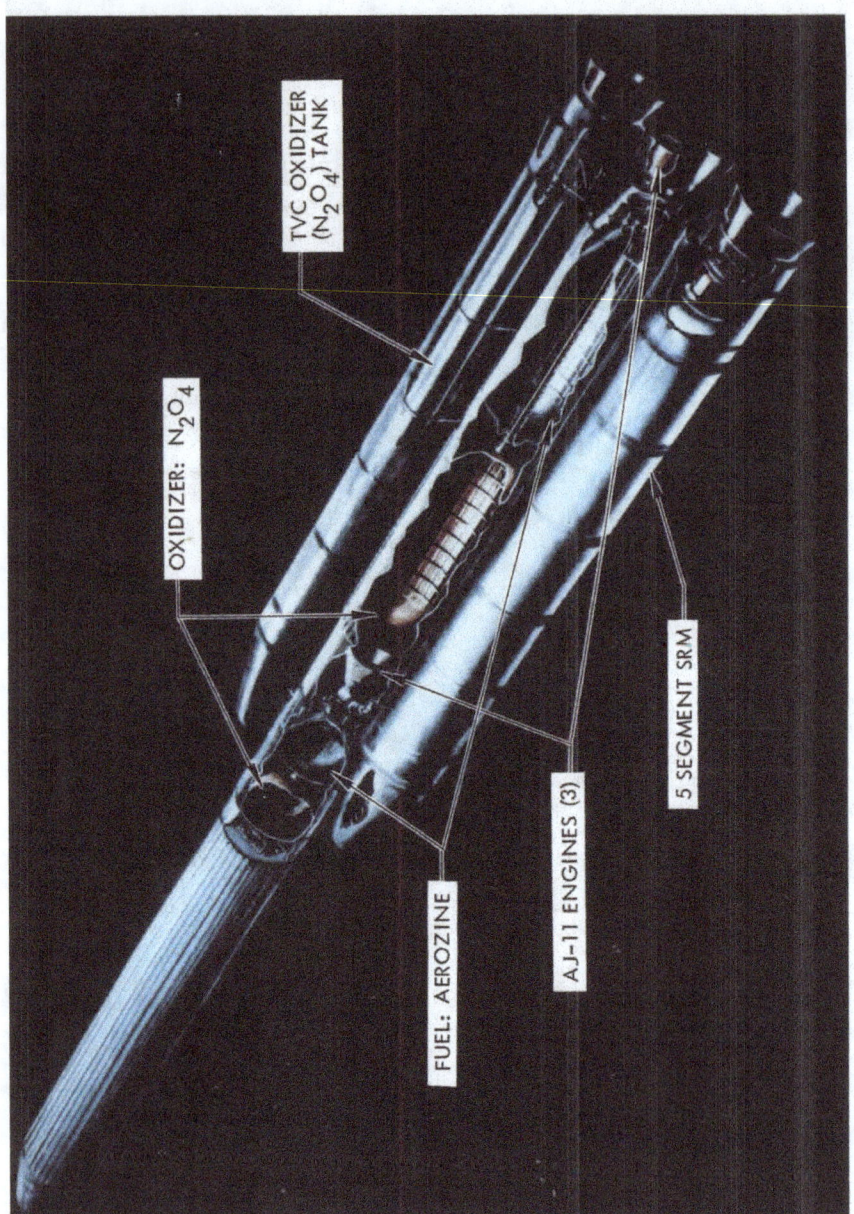

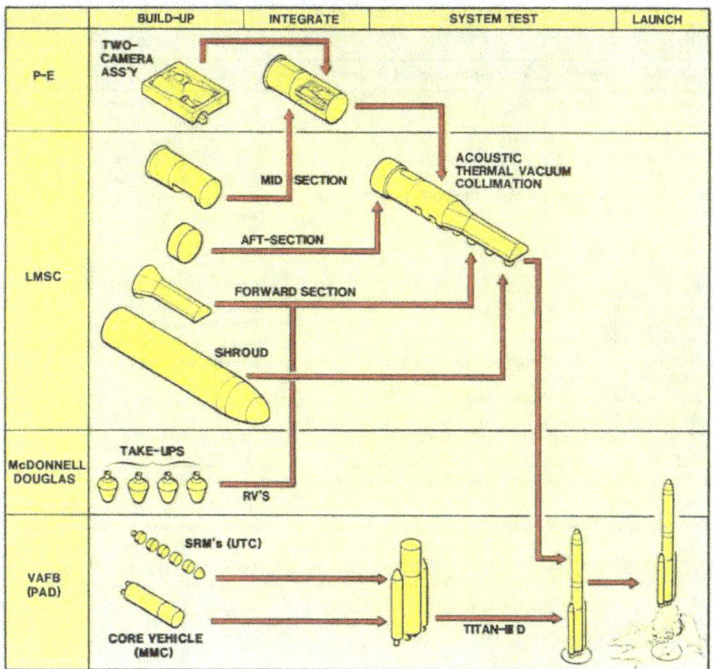

Factory-to-Pad Process

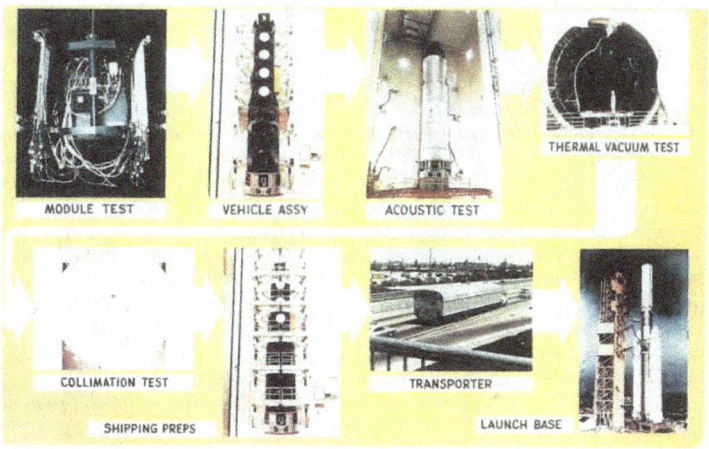

Hardware Flow

Table

Program Personnel* during Development Phase

System Program Director (SAFSP)	Maj. Gen. John Martin
Director of Special Projects (CIA)	John J. Crowley
System Program Office (SAFSP)	Col. Frank S. Buzard
Sensor Subsystem Program Office (CIA)	[redacted]
Sensor Subsystem Program Mgr. (Perkin-Elmer)	Michael Maguire
Satellite Vehicle Program Mgr. (LMSC)	Stanley I. Weiss
	Paul Heran
Reentry Vehicle Program Mgr. (McDonnell-Douglas)	Logan MacMillan
Stellar-Terrain Camera Program Mgr. (Itek)	John Doyle
Extended Command System Program Mgr. (GE-AESD, Utica)	Norman N. Feldman
T'Unity Software (TRW)	Robert M. Larkin
	Winston W. Royce
Mark-V Reentry Vehicle Program Mgr. (GE Reentry System Div)	Stephen Csencsitz
System Engineering Contractor Mgr. (Aerospace Corp.)	Leonard C. Lidstrom
System Engineering/Technical Support Mgr. (TRW)	C. W. Besserer

*See Appendix F for a listing of personnel for the entire program.

Development Progress

By the end of 1967, the HEXAGON program had made some progress toward a first launching date of April 1970. The general vehicle configuration—Titan-IIID booster with five segment solids, a satellite vehicle 120 inches in diameter with four RVs—had been defined. Although Perkin-Elmer had been working for 15 months on the sensor subsystem, it was progressing slowly; the preliminary design review of the sensor subsystem, scheduled for June1967, slipped to December and would eventually take place in January 1968. The system critical design review, scheduled for October 1967, then slipped to October 1968. LMSC was now on contract for the SBA including design of the aft (control) section of the vehicle and work on subsystems was progressing. Interfaces between the contractors were being negotiated and, by the end of the year, such items as electrical power voltage levels (22-32 volts vs. 25-33 volts), wire gauge (22 vs. 26), connector sizes, and film supply-reel orientation had been resolved. The midsection, which was to house the sensor subsystem, was being redesigned. This redesign was caused by the fact that, until sensor/SBA detailed interfaces were resolved, the SBA contractor, LMSC, had no detailed design requirements in this area. In both the SBA RFP and the subsequent general specification (DS 10001) it was merely stated that "the SBA structure external to the sensor subsystem shall orient, protect, and support the sensor subsystem . . . . Sensor subsystem dimensions shall be such that a section of the satellite vehicle, 10 feet in diameter and 180 inches in length, will house all the equipment . . . ."[200] There appears

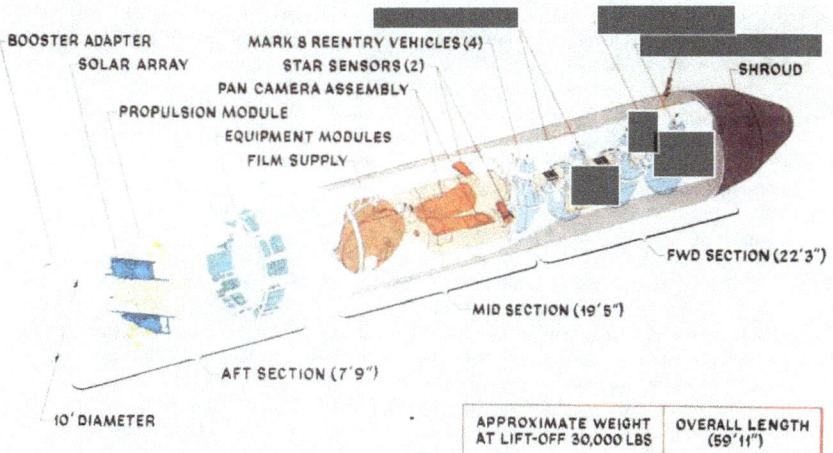

Satellite Vehicle Configuration

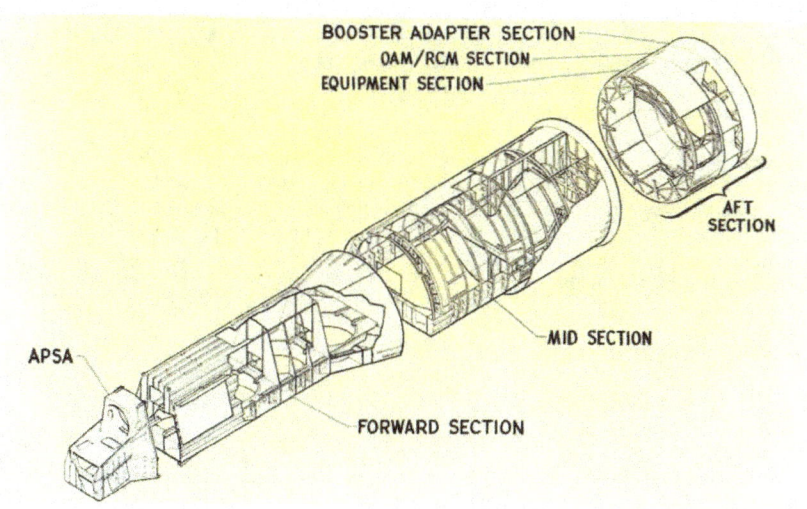

Satellite Basic Assembly Structure

to have been confusion in this important area, as the SSPO interface document issued by the SETS contractor (TRW) as late as 31 July 1967 instructed Perkin-Elmer that the available space for the sensor was a diameter of 90 inches and a length of 170 inches (vs. 120 inches and 180 inches, respectively, used by the SPO). Inasmuch as the SSPO had access to the SBA RFP this discrepancy is hard to explain. In his draft history of the program[200] ▮▮▮▮▮ holds that these changes (90 to 120 inches diameter and 170 to 180 inches length) were made by LMSC between the time of source selection and contract award and claims this had a cost impact in the sensor. A comparison of before and after LMSC drawings does not support this view, even though some changes in external structural concept were made.

During the first half of 1968, the SPO, SSPO, and the respective contractors—LMSC for the SBA and Perkin-Elmer for the sensor subsystem—resolved major differences and began to publish integrated plans, schedules, and technical interface documents. There was agreement on the total flow of equipment from each contractor's factory to the integration location; testing, including sensor operation, was to be performed at LMSC and at VAFB. There was also agreement that, if these activities were accomplished on schedule, a first launching date of 1 October 1970 could still be met.

System Description

With all components on contract, and system design practically fixed, the HEXAGON system was becoming well defined. The orbiting vehicle would be 10 feet in diameter and 52 feet in length. It would be made of three sections: forward, mid, and aft. The forward section, built by LMSC, would be 27 feet 9.3 inches long and would contain the mapping-camera module (Itek and GE), the four RVs (McDonnell-Douglas) with film take-up, and the forward film-path (Perkin-Elmer). The midsection, built by LMSC, would be 19 feet long 6 inches, and would contain the sensor subsystem (Perkin-Elmer). The aft section, also developed by LMSC and referred to as the satellite control section (SCS) would be five feet long and would contain the controls for the various satellite subsystems plus the booster adapter for mating to the Titan-IIID booster.

The Satellite-Control Section (SCS)

The SCS contained all the subsystems concerned with the operation of lthe satellite vehicle on orbit.

a. The attitude-reference module (ARM): sensors, gyros, and electronics to determine the attitude of the satellite.

b. The reaction-control module (RCM): the small hydrazine monopropellant thrusters and associated tankage and plumbing to maintain the satellite in the proper attitude.

c. The orbit-adjust module (OAM): the large hydrazine monopropellant engine and associated tankage and plumbing to provide the impulse to maintain the satellite in the proper orbit, to change the orbit of the satellite when necessary, and to deorbit the satellite after the mission was completed.

d. The solar-array module (SAM): the solar wings and associated electronics to charge and recharge the main batteries.

e. The electrical-power module (EPM): the main batteries, originally seven, to provide the power for all the satellite and payload functions. These batteries were kept charged by current from the solar arrays. In addition, the charge controllers and thermal cutoff relays were part of this module.

f. The tracking telemetry and command module (TTCM): the transmitters, receivers, recorders, telemetry equipment, and the extended command system, which was the "brains" of the system, receiving and storing commands from the ground stations and transmitting vehicle-status information to them. The minimal command system, which provided a limited command capability to operate the satellite in the event of malfunction in the extended command system, was also a part of this module.

g. The back-up recovery attitude-control system (BRAC or Lifeboat [202]: the emergency attitude control system and command system to provide a means to position the satellite for recovery or reentry if the primary attitude-control module, the reaction-control module, or the extended command system became inoperative.

Most of the modules had extensive redundancy and "cross-strapping" capabilities which enabled the ground controllers to switch the connections between different elements in the event of malfunction of some primary element. For example, the propellant tanks of the orbit-adjust engine and the reaction-control thrusters could be connected to feed either the orbit-adjust engine or the small reaction-control thrusters. Similarly, the redundant gyro in the attitude-reference module could be connected to bypass failed electronic components.

The Sensor Subsystem[203]

The sensor subsystem consisted of the two camera assemblies, the film supply, and four take-ups. The sensor subsystem two-camera assembly, located in the satellite vehicle midsection, contained a pair of panoramic cameras mounted in a frame. One camera looked forward on the satellite vehicle (Camera A, port side) and the other looked aft (Camera B, starboard side). Each camera had a 60-inch focal length, f/3.0 folded Wright optical system. The optical system, which contained both reflection and refracting optical elements, was mounted in the optical bar.

The cameras could be operated in any of 16 scan modes (30 to 120 degrees with center angles 0 to ±45 degrees) as selected by the "T'Unity" software, with frame format length determined by the scan mode in use. Scan modes were selected as an in-flight option on a per-operation basis. The selected mode remained constant throughout that operation, giving Mission Control a maximum target coverage capability with minimum film wastage. (The original sensor subsystem design had only a 120-degree scan width. An NRO study, completed in December 1966, recommended including scan widths of 30, 60, and 90 degrees, with variable scan centers of 15, 30, 45, and 60 degrees off nadir.)

During photography, the optical bars rotated continuously through 360 degrees to provide cross-track scanning, although photography occurred only during a maximum of 120 degrees of scan. In each optical bar, a platen (directing the film across the focal plane) was electronically locked to the optical bar through 130 degrees of scan (120 degrees scan plus 10 degrees for settling time, corresponding to the maximum cross-track coverage for the available scan modes) and then recycled to the start-of-scan position.

Characteristics of the HEXAGON Search and Surveillance Sensor

Optics	60-in. focal length, f/3.0 folded Wright (modified Schmidt) system (T ¾ excluding filter factor)
Aperture diameter	20 in.
Field angle	±2.85°
Slit width range	0.91 in. (maximum); 0.08 in. (minimum)
Film	6.6-in.-wide (black and white) Type 1414, SO-208, and others. Also, SO-130 (infrared false color) and SO-255 (natural color).
Resolution (2:1 contrast)	Center of format 200 l/mm; elsewhere in format 160 l/mm
Film load	Initial load 104,000 ft. of 6.6-in. film/camera. Ultimately 155,000 ft./camera
Film stack diameter	68 in.
Scan modes	30°, 60°, 90°, and 120°
Center of scan	0°, ±15°, ±30°, and ±45°
Maximum scan angle	±60°
Stereo convergence angle	20°
Frame format (120° scan)	6-in. by 125-in.
Film velocity	200 in./sec (maximum) at focal plane
Image motion compensation range	0.018 rad/sec to 0.054 rad/sec for Vx/H, ±0.0033 rad/sec for Vy/H
Weight (less film)	5,375 lbs.

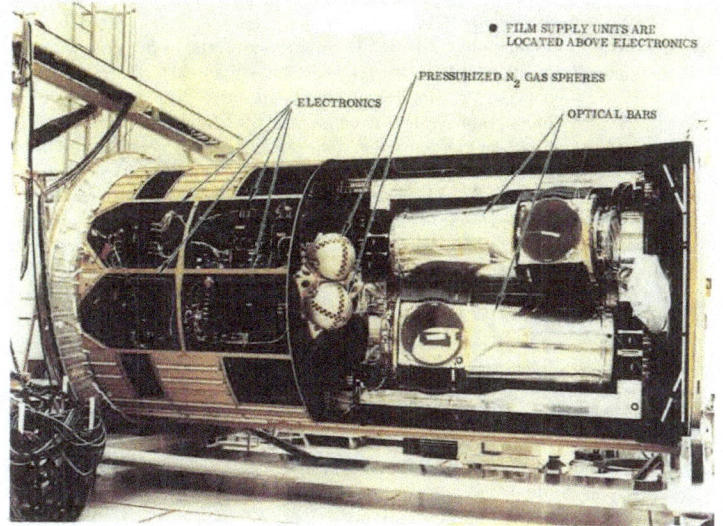

Optical Bar Panoramic Camera Installation

Satellite Control Section

The sensor subsystem was organized into units so that most interactions occurred within the units; individual units interacted as little as possible with each other. The sensor subsystem electronic and electro-mechanical modules were either installed in the electronics compartment—mounted on the two-camera frame—or integrated with subassemblies. The distance the film traveled, from the supply assembly in the aft section to the first RV in the forward section, was approximately 140 feet (in both cameras). Throughout its travel over 124 rollers in camera A, 131 rollers in the camera B, and six airbars in each camera, the film was to remain centered within specified tolerances. To correct the displacements of supporting film-path elements (such as rollers and air bars) caused by structural deformations due to launching and thermal variations, each camera contained active and passive articulators to steer the film at critical points in the film path. Active articulators also steered the film across the sensor subsystem primary bulkheads (that is, between the supply and the midsection and between the midsection and the forward section) to prevent the film from telescoping on the supply and take-up cores. Passive articulators maintained film-path alignment between the RVs and across the two-camera assembly frame in each film path.

The supply assembly maintained film-stack integrity in all conditions of powered flight and orbital operation. It supplied film to the two-camera assembly at controlled constant velocities up to 70 inches per second. Each take-up assembly—one in each of the four RVS—had a film capacity of one-fourth the film load of the supply assembly. Film was pulled from the camera looper by the take-up drive motor and core. The looper assembly in each film path served as the interface between the coarse and fine film-transport systems. In the fine film-transport system, the film was accelerated to 200 inches per second, decelerated, and recycled, while the platen cycled through the photo-recycle phases.

The looper allowed the total length of stored film in it to be constant, but the relative film lengths in the supply and take-up sides of the looper varied as a function of looper carriage position. The twister assembly, located in the film-drive assembly, accommodated the angular change between the film-drive assembly rollers (which were fixed to the frame) and the rollers in the platen assembly (which were locked to the optical bar during the photographic cycle). The twister assembly consisted of a twin air-bar assembly and a housing that incorporated a manifold through which nitrogen gas was supplied to the air bars. The film wrapped one of the air bars before wrapping the entrance roller of the platen assembly and wrapped the other air bar after leaving the exit roller of the platen assembly. The twister assembly was free to rotate about its pivot-point in response to angular changes between the rollers in the film drive assembly and those in the platen assembly.

The film was completely enclosed in light-tight, pressurized assemblies throughout its passage from the supply assembly to the take-up assembly. The film, as loaded in the supply assembly prior to launching, contained approximately 65 pounds of water, providing an effective relative humidity of approximately 40 percent at ambient temperature. The enclosed pressurized film-path prevented rapid vaporization of the water from the film emulsion during system operation. Excessive vaporization could cause two harmful effects: (1) flatness distortion of the film, making it difficult to track and producing flutter in the focal plane, and (2) creation of a gas layer between film wraps in the take-up assembly, causing uncontrolled telescoping as the stack built up.

The primary (two spherical tanks) and supplemental (one spherical tank) pneumatics systems supplied dry nitrogen gas to pressurize the sensor subsystem's enclosed film path. (Each of the active film steerers contained nitrogen airbars to preclude damage to the film.) These bars contributed to pressuring the film path. These systems contained approximately 109 pounds of nitrogen under a nominal pressure of 3,265 psia at 70 degrees Farenheit.

The supply assembly supported, protected, and drove the film supply for both the forward-looking and aft-looking cameras. Initially in the program each supply reel carried 104,000 feet[204] of 6.6-inch-wide Type-1414 film and weighed 890 pounds. The two-camera assembly and the supply assembly were mounted in the midsection of the satellite vehicle.

The Mapping-Camera Module

The mapping-camera module contained the stellar-terrain camera and its light baffles, electronics, film paths, and thermal controls; the doppler beacon and antenna; the Mark-V RV; and the structure to support all of these items.

The terrain camera had a 12-inch f/6.0 metric lens with eight elements. It used 9.5-inch film. The stellar camera, which imaged stars above sixth magnitude, had two 10-inch f/2.0 lens systems—one looking out each side of the module. It used 70-mm film.

The RV was an improved version of the Mark-V vehicle, originally developed for the CORONA program, modified to accommodate the 9.5-inch and 70-mm film take-ups.

The doppler beacon assembly provided data for more accurate determination of the vehicle orbit.

The entire module was assembled and tested at Itek, then shipped to LMSC for integration with the rest of the HEXAGON system and final systems testing.[205]

The Donovan Review Committee

In October 1968, Maj. Gen. John Martin of SAFSP became concerned that divided management responsibilities and the general complexity of the HEXAGON program might lead to inadvertent omissions or errors in design. He asked Dr. Allen Donovan, senior vice president/technical of Aerospace Corporation, to convene a committee of senior aerospace experts to conduct a "general system engineering review"[206] of the entire program.

Mark 8 Reentry Vehicle

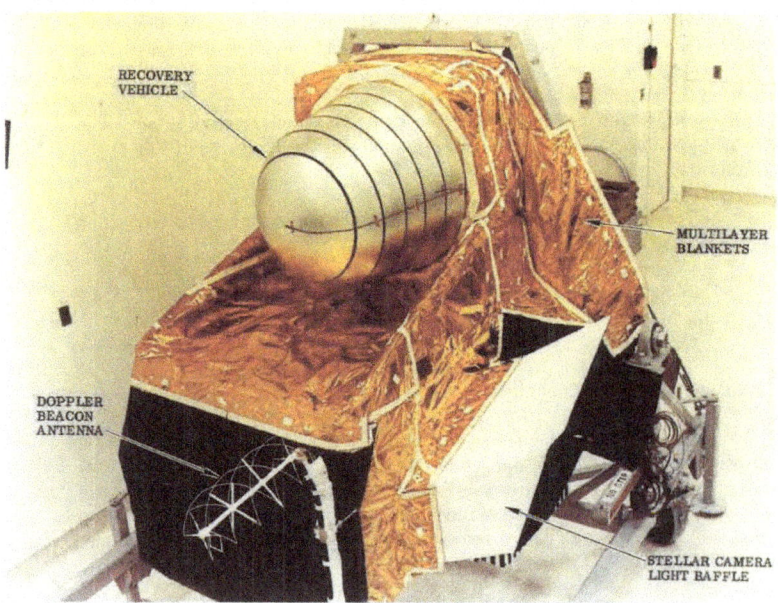

Mapping Camera Module

SECRET
NOFORN-ORCON

After visiting all contractors and meeting the managers and technical people at each plant, the committee concluded that, while the hardware program was not optimal, it was generally satisfactory. Their major conclusions were the following:

- The current passive thermal control system was not adequate; an active control subsystem was recommended.

- The electrical power system was marginal and should be augmented.

- Development of the operational control software, which was not yet on contract, should be undertaken immediately.

Thermal Control

The design of the thermal control subsystem had been a major problem from the start. Since HEXAGON would remain on orbit for 36 to 60 days, an "active" system with heaters and thermostats (as used on previous systems) would be a major power consumer. A "passive" system would be one in which the temperature within the satellite was maintained at the proper value (70 degrees Farenheit ± 23 degrees) by using a specially designed paint pattern on the satellite's surface. This paint design would control absorption of energy from the sun as well as the satellite's radiation of energy back to space, thus maintaining a proper temperature. The design of paint patterns depended not only on characteristics of paints themselves but also on the amount of heat generated inside the satellite by electrical motors, sensor electronics, and other equipment.

LMSC and Perkin-Elmer each believed it was better qualified to do a thermal paint design and, therefore, should have design responsibility. The two Government offices supported their respective contractors, and this disagreement and resultant debate lasted through 1967 into 1968. The matter was finally resolved in February 1968, when LMSC was given responsibility for design and Perkin-Elmer was directed to review LMSC's work.

Later Perkin-Elmer became concerned about the effects of humidity on the pressurized film path, since tests showed that under certain conditions film would stick to the rollers and air bars, disrupting smooth transport. Because no one had ever flown 1,576 pounds of film (two rolls 66 inches in diameter), very little was known about how such a mass might behave in a space environment. Studies were undertaken to determine the water content and the outgassing characteristics of the film. Eastman Kodak was asked to supply film with a relative humidity of 40 percent rather than the normal 45 percent ±5 percent. Concern over the problem persisted; eventually, it was decided to control the temperature gradient along the film path to ± 3 degrees—a requirement that was completely beyond the capability of the passive control system. As a result, in 1969 it was decided to install an active system—made up of thermostats, heaters, controllers, and multilayer thermal insulation—along the film path. This arrangement increased the power consumption of the system, so two solar panels were added to the 20 already planned.

SECRET
Handle via
BYEMAN-TALENT-KEYHOLE
Control Systems Jointly
BYE 140003-92

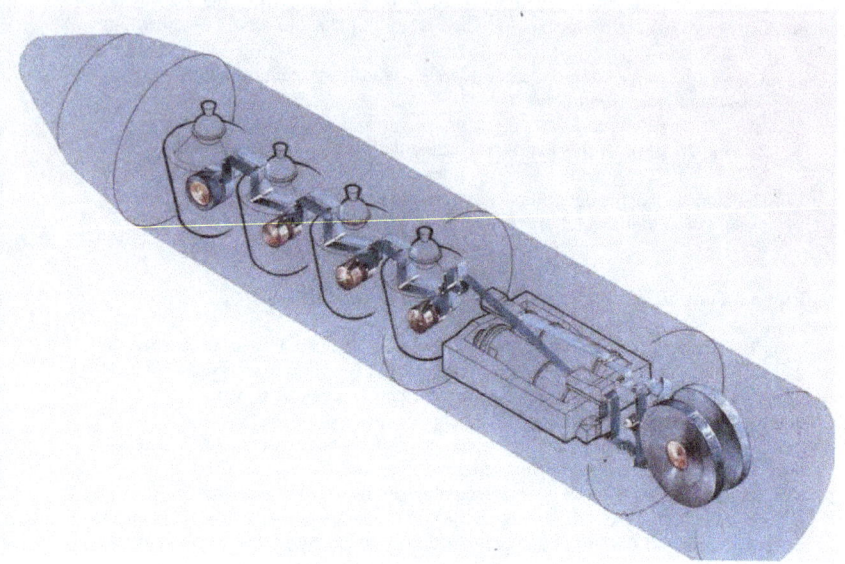

Film Transport

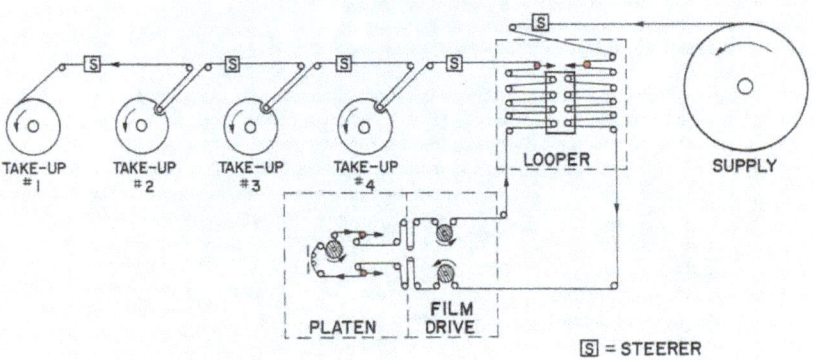

Film Path Schematic

Operational-Control Software

Development of on-orbit operational-control software for the system was the final major issue between the CIA and SAFSP. The CIA desired to control the satellite from the Satellite Operations Center (SOC) in Washington, sending specific command instructions for sensor operations to the STC for re-transmission to the satellite (as had been done on the CORONA program). SAFSP, on the other hand, maintained that the complexity of the system—including the sensor subsystem—required that all control of the satellite be done by the Satellite Control Center (SCC) at Sunnyvale, California. It was decided that the SOC would send the list of requirements (targets and target areas), with their priorities, to the SCC where actual target selection for a particular revolution would be made (considering weather conditions and vehicle health) and sent as a command message to the satellite.[207] TRW won the competition to produce appropriate software, called "T'Unity." Even though this was the last part of the system to go on contract, it was not a pacing item in the program.

By the summer of 1969, it was clear that the projected launching date, which had already slipped from October 1970 to December 1970, was still in jeopardy. All major contractors were behind schedule.

Development Problems

At McDonnell-Douglas (the RV contractor) the development of the parachute system (which had been subcontracted to Goodyear Aerospace) was in serious difficulty. The drogue, which was to pull the main chute from the pack, could not do so. The main chute was completely destroyed in seven consecutive tests; it was too weak to sustain the forces it experienced during deployment. In addition, the chute was unstable, oscillating from side to side as much as ± 32 degrees. (The equivalent figure for the CORONA chute was ± 7 degrees.) This oscillation made it almost impossible for recovery pilots to make aerial pick-ups. A number of "band-aid" fixes were made to remedy the problem: vent holes were placed in the canopy to prevent severe oscillation, three belly-bands of stronger materials were sewn around the canopy to strengthen it, and a standard drogue chute was tested as a replacement for an unsatisfactory "ballute."

GE, which was responsible for the extended-command system (the "brains" of HEXAGON), was far behind schedule because of parts shortages and design problems. At LMSC, gyro problems had developed in the attitude-control subsystem. In addition, the design of the newly required active thermal-control system was behind schedule. Perkin-Elmer had continuing problems with the film-transport system: the film mis-tracked, ran off the rollers, and jammed the system (as well as other parts of the sensor).[208]

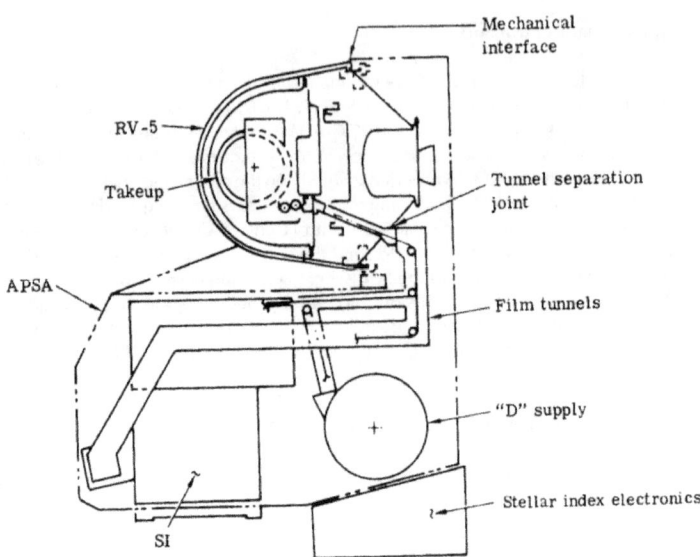

Mapping Camera Module

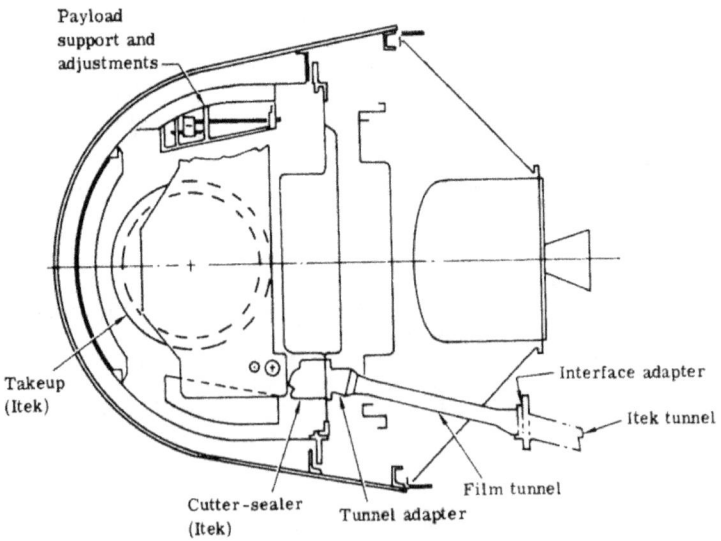

SRV Inboard Profile

Mark V Reentry Vehicle

Section 10

The Hexagon Flight Program

The DNRO and the Intelligence Community were concerned that further slips in the launching schedule might result in a period during which there would be no photocoverage of the USSR. In 1967, nine CORONAs of the J Series had been launched; in 1968, eight more Js had been used. By 1969, there were only 14 CORONAs remaining in the inventory. Should more CORONAs be procured? If so, how many? And what should be done about HEXAGON, which was continuing to experience development difficulties at all of the major contractors?[209]

In addition, almost from the start of the HEXAGON program there had been critics who maintained that the system's requirements could be satisfied less expensively by improving CORONA or by using some other less sophisticated system. When the costs of HEXAGON at Perkin-Elmer alone began to rise from the proposed ▆▆▆▆ as of September 1966 to an eventual negotiated cost of ▆▆▆▆ in February 1968 (and an actual final cost of ▆▆▆▆),[210] and the other contractors were beginning to show similiar cost increases, these efforts were intensified. In the spring of 1969, the Bureau of the Budget (BoB) convinced the new President, Richard Nixon, that the HEXAGON program should be canceled, since it could be replaced by a combination of improved CORONA and GAMBIT satellites. This provoked an immediate response from the CIA and others in the Intelligence Community who saw a strong need for HEXAGON capabilities. On 15 June 1969, the BoB decision was reversed and HEXAGON was reinstated. In November 1969, BoB made one more effort to cancel the program but there was general agreement that, with the SALT negotiations underway, HEXAGON was more needed than ever.[211]

These program perturbations caused some concern to the various contractors and the program offices but, in general, had little effect on progress with HEXAGON.

About this time, DNRO McLucas gave his deputy, Dr. Robert Naka, the task of conducting an independent study of the HEXAGON schedule specifically to determine how the remaining CORONAs should be used. Naka, meeting with Col. L. S. Norman of SAFSP and ▆▆▆▆ of the CIA, concluded that there was a 95-percent probability of a HEXAGON launching not later than June 1971, a 75-percent probability for March 1971, and a 50-percent probability of meeting the current official schedule of December 1970. They also concluded that there was a 95-percent probability that one of the first three HEXAGONs would be successful. On this basis, Naka suggested that the 12 remaining CORONAs be rescheduled so that at least two launchings could take place after July 1971.[212] Naka's committee met again in October 1969 and January 1970 to review program progress and to reassess the need for rescheduling CORONAs (or for ordering additional CORONAs); it did not change the original conclusions.[213]

During the first and second quarter of 1970, there was a continual juggling of test hardware, test plans, and schedules in an effort to maintain the December 1970 launching date. As black boxes and subsystems became available, they were placed in the satellite developmental vehicles (SDVs), which were then used to check out all test procedures, ground-test facilities, thermal-vacuum test chambers, and the launching facilities at Vandenberg. This system testing was a crucial phase of the program—the proving ground for all hardware from the associate contractors.

DNRO John L.
McLUCAS

While this development activity was going on, the first items for the first flight were being assembled and tested at the respective contractor plants. At Perkin-Elmer, the flight-sensor subsystem was being installed in the midsection; at LMSC, the forward section and aft section build-up was under way. Several problems persisted and, before long, the first launching date had to be slipped from December 1970 to March 1971. Another setback occurred on 7 July 1970 when the first flight-article camera assembly suffered a catastrophic failure while undergoing testing at the Perkin-Elmer plant. It was necessary that the second flight-article sensor be substituted for the first HEXAGON flight.

The first flight vehicle (SV-1) was assembled at LMSC in December 1970 and the system test sequence began. Acoustic tests (to simulate the ascent environment) and thermal vacuum tests (to simulate the orbital environment) were generally successful. These tests took much longer than planned; before long, a March launching was out of the question and the date slipped—first to April 1971 and later to "not earlier than" 2 May 1971. In late April, while final preparations for shipping the SV-1 were under way, a shutter assembly failed during extended service-life testing. The decision to remove, inspect, and replace the shutter assembly meant that first launching would slip to mid-June, since the entire reassembled vehicle had to go through additional collimation testing to verify camera performance.

When SV-1 was finally ready for shipment to VAFB, a more mundane problem arose. The State of California restricted use of the SV-1 transporter (a mammoth vehicle some 14 feet high, 14 feet wide, and about 70 feet long) to daylight, weekday, and non-rush hours. It was now 28 May—the Memorial Day weekend—and movement of the satellite had to wait until after the holiday. But once the SV-1 arrived at VAFB, things began to go smoothly. All prelaunching tests and preparations were completed without incident. At 1141 PDT, 15 June, the first HEXAGON vehicle was launched into orbit—noisily and successfully.

Initial on-orbit tests showed that all subsystems were operating normally. The vehicle was stable, the solar panels were deployed, the command and telemetry subsystems received and transmitted data, and the sensor was working. But about 8 to 10 hours after launching, it became apparent that temperatures in the main battery bay—particularly on batteries 3 and 4—which should have stabilized between 35 and 75 degrees Farenheit were actually as high as 80 degrees and continuing to rise. The cause of this problem was not known; but it was feared that at about 135 degrees these batteries would explode into shrapnel, producing catastrophic results. Fortunately, during the night the battery temperature stabilized with a cycle between 88 and 100 degrees.

But another problem arose. Since the temperature of the batteries would rise when they were being charged by current from the solar panels (and also when they were being discharged to operate the satellite), the power system had been designed with thermal relays, which would open at about 100 degrees, cutting off the charging current. When the batteries cooled, the relays would close and charging would begin again. If this cycling permitted an adequate charge to build up in the batteries, the mission could continue in essentially a normal manner; however, if the batteries became too completely discharged, they could not be recharged by the solar panels and would degrade, in a short time, to a point where the vehicle could not operate. There was, on this flight only, a reserve main battery, with sufficient capability to operate the vehicle for four or five days to help ensure some photography on the flight; however, once the switch had been thrown to the reserve main battery, it could not be returned to the main supply.

During the morning and early afternoon of 16 June there were numerous teams of contractor, SPO, and SSPO personnel collecting data, studying schematics, developing alternatives, and trying to decide on a course of action. Schemes were developed for reducing the power load, such as restricting payload operation to only a few (4-4.5) minutes per revolution and switching other power consumers off. A final decision could not be delayed much beyond 1600 PDT, because after that time there would be no opportunity to command a change before the batteries expired.

At the decision meeting (1600 PDT, 16 June), the Aerospace Corporation and most contractor advisory personnel were in favor of the less risky option: switching to reserve supply (getting one RV of photo-material but giving up all chance for a more productive mission). A few brave souls, led by Buzard—who made the final decision—opted for another solution: continue on main supply. Buzard's boss, SAFSP Director Brig. Gen. Lew Allen, Jr., who had recently replaced Brig. Gen. William G. King, Jr., (1969-71), supported him in this choice, and the mission continued on main batteries.

USAF Brig. Gen. Lew ALLEN, JR.

Brig. Gen G. William KING, JR.

In subsequent days, as the problem became better understood, the operating team kept a very careful account of battery voltage and power available and scheduled operations accordingly. The sensor operating time gradually increased from 17 minutes per four-revolution span to 30 minutes, which, while only about one-half the design capability, did not limit the photographic task appreciably. Once past this hurdle, the entire vehicle operated with only minor problems. The sensor subsystem transported 40,000 feet of film into RV-1 which, while not a full load, was near the limit originally set for that capsule.

On 20 June 1971, during orbital revolution 82, the first RV was separated from the satellite and reentered in the Hawaiian recovery area. Recovery forces sighted the capsule, but the parachute was so badly damaged that aerial recovery was not attempted. The capsule landed in the water, where it was retrieved by surface forces, and taken to Hickam AFB for transport to the processing laboratory at Eastman Kodak in Rochester, New York. While the primary objective of the HEXAGON mission was to provide high resolution photography over broad areas, the intent of the first flight was to demonstrate functional operation of the system. The sensor system certainly achieved this intent.

One of the NPIC representatives at the Eastman Kodak processing facility remarked, "My God, we never dreamed there would be this much, this good! We'll have to revamp our entire operation to handle the stuff."[214]

Between revolution 82 and revolution 179, operations were routine and normal. Based on the analysis of film from RV-1, numerous commands were sent to adjust the camera for better performance. The operations team developed procedures for tracking the battery voltage to determine how much power was available for payload (and other) operations. The limitation of 30 minutes of payload operation per four-revolution cycle imposed no constraint on the general operation, and 52,000 feet of film was moved into RV-2, which was recovered on revolution 179 on 26 June. This time, parachute damage was less severe and aerial recovery was successful.

On-orbit operations were generally routine from revolution 179 through revolution 405. Despite an emergency shutdown of the sensor subsystem on revolution 314, film moved into RV-3. Unfortunately, on 10 July during the recovery attempt, the parachute was completely destroyed and the capsule sank on impact.

As a result of parachute problems on RV-1 and RV-2 and the loss of RV-3, a limit of 50 percent of load (26,000 feet of film) was placed on RV-4. By this time, both the operations teams and the satellite were tired. There were more emergency shutdowns of the sensor subsystem, presumably caused by film-path problems. In addition, the attitude-control thrusters began to degrade and usage of attitude-control propellant increased. On revolution 484, the voltage on the pyro batteries—essential to the recovery sequence—began to drop alarmingly, indicating that they were nearing depletion and that early recovery was desirable. On 16 July, during revolution 502, RV-4 with 26,000 feet of film onboard, reentered and was successfully caught by one of the recovery force's C-130s.[215]

The operations team continued to command the HEXAGON vehicle, exercising the various subsystems, conducting experiments on the attitude-control system, the orbit-adjust system, and Lifeboat (the back-up recovery control system). On 6 August 1971, after 52 days on orbit, SV-1 was deboosted into the Pacific Ocean. During its active phase of 31 days, it had transported 175,601 feet (1,350 pounds) of film and conducted 430 photo-operations at an average ground resolution of 3.5 feet and a best Controlled Optical Range Network (CORN) target resolution of 2.3 feet.[216] Of this 175,601 feet of film, 123,601 feet (930 pounds) had been recovered in three RVs.[217]

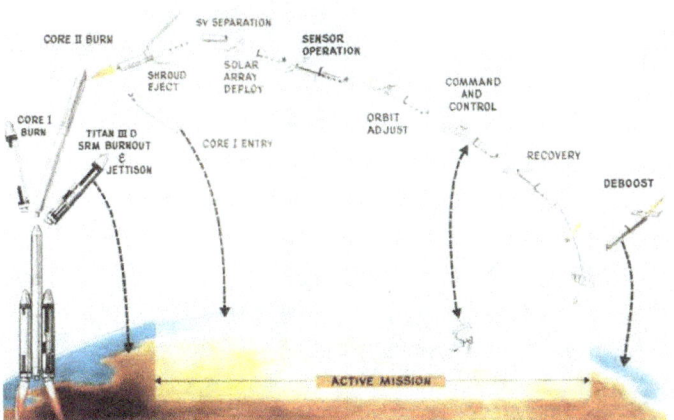

Ascent, On-Orbit, Deorbit Sequence

Aerial Recovery By C-130

Satellite Vehicle Assembly Going Vertical

Satellite Vehicle in Test Facility

As a basis for comparison, the first successful CORONA recovery (August 1960) carried 20 pounds of film. Later CORONA versions carried 40 pounds; the double-capsule version had 80 pounds. In the GAMBIT program, GAMBIT-1 had 45 pounds of film; in GAMBIT-3 the double-bucket carried 160 pounds.

On 24 June 1971, two days before the successful recovery of RV-2 from mission 1201 and the completion of mission segment 1201-2, Colonel Buzard left the HEXAGON Program Office, having been program director from program inception through all of the difficult days of program definition, source selection, interface resolution, complex development, schedule and performance pressure, and exacting testing. His outstanding leadership and devotion had been rewarded by a very successful first flight. He was assigned to duty with Gen. Allen as his Vice Director; he retired from the Air Force on 1 November 1972. Col. Robert H. Krumpe succeeded Buzard as HEXAGON Program Director on 24 June 1971.

As a result of the SV-1 experience, a number of items had to be improved before the next launching: the parachute system, the pyro battery, the battery bay temperature, and the attitudes control thrusters.

The parachute system, which had been subcontracted to Goodyear Aerospace, needed to be completely redesigned. McDonnell-Douglas and Henry Epple (of Aerospace Corporation) designed a new, stronger, more stable "extended skirt" chute which was manufactured by Para Dynamics, Inc., of El Monte, California; the new drogue chute was designed by Irving Air Chute Company. This new design was carefully tested and proved vastly superior to the Goodyear version; it was used on all subsequent flights. In order to ensure the quality of the chute, Epple and a McDonnell-Douglas representative inspected the completed chutes and personally packed them, using a vacuum technique to extract air and reduce volume. The chutes were "baked" at 370 degrees for 8 hours to set their shape; then they were installed in the RVs.[218]

Controlling the battery temperature was a more difficult problem, since the cause of the difficulty was not known. There were several theories:

- The aft section was absorbing more solar energy than predicted.

- The aft section thermal-control surfaces were improperly applied or were damaged before launching.

- There was a basic design error.

- Ascent events caused contamination of the thermal-control surfaces.

After analysis and a ground-testing program eliminated the first three theorized causes, efforts were concentrated on ascent contamination. Four SV-2 sensors were placed on the battery bay of the second satellite vehicle to determine precisely when contamination occurred. Was it from the cloud of dust at solid rocket motor (SRM) ignition and liftout? Was it from the small solid rockets that staged the SRM after burnout? In the meantime, something had to be done about the batteries in SV-2. Fortunately, there was unused space on the other side (the cool side) of the vehicle, and the batteries were moved to that location.

USAF Col. Robert H. KRUMPE

All of these fixes, plus the normal testing sequence, slipped the launching of SV-2 to 21 December 1971. Then, during the countdown, a problem in a Titan pyro circuit caused an additional delay, while the suspect wiring harness was replaced. As a result of this delay the thrust-vector control valves on the SRMs corroded and were replaced.

The second HEXAGON vehicle (SV-2), mission 1202, finally was launched into its planned 86- by 189-nm orbit on 20 January 1972. Although there were some problems, photographic operations took place on day 1 through day 39 of the mission.

During the latter part of RV-2 operations and at 43 percent of clocktime through the mission, there was a catastrophic failure of the forward-looking camera in the sensor subsystem: the film was broken during a camera operation.[219] The rest of the mission was limited to monoscopic coverage only, using the aft-looking camera. The mission was troubled further by excessive propellant usage in the satellite vehicle's reaction-control subsystem (RCS); this subsystem, like the orbit-adjust subsystem [OAS], used monopropellant hydrazine thrusters). The propellant tankage of the RCS and OAS was cross-strapped so that excessive RCS propellant needs were largely accommodated by using propellant nominally planned for OAS usage. Despite the use of a redundant set of RCS thrusters, the rate of RCS propellant usage was enough above nominal that the mission was terminated during revolution 632 on day 40 using the Lifeboat-II[220] subsystem, with no solo[221] phase.

As a result of the battery problems experienced on the first HEXAGON flight, the batteries in bay 12 had been moved to bay 3 to prevent overheating. This change proved effective on SV-2. Also, SV-2 was instrumented with quartz crystal microbalances and calorimeters to determine the cause of the problems seen on the first flight. This instrumentation showed that the solid-rocket staging event was the source of contamination of the thermal-control surfaces.

Approximately 100,000 feet of film was recovered from the A-side camera in the four RVs and about 56,000 feet from the B-side in RVs 1 and 2. Resolution "was degraded by the need to use larger slits to compensate for the low sun angles, scene characteristics (snow and blowing snow), and ground haze typical of the winter season."[222] Because of parachute damage during Mission 1201, all main chutes were modified, as described above, and deployment was delayed until the RV reached 40,000 feet. This modified design performed satisfactorily on mission 1202; all four RVs were recovered aerially.

During preflight planning for the third HEXAGON vehicle, mission 1203, a principal concern was managing an anticipated RCS thruster problem. Previous flight data and ground tests "indicated that thruster-valve leakage and subsequent degradation was caused by particulate deposits on the thruster-valve seats."[223] Possible sources of these particulates were contaminated fuel and non-volatile residue (NVR) building up in the RCS tanks after they were filled with propellant. Test and analysis showed that residues resulted from exposure of the hydrazine propellant to the rubber diaphragm in the RCS tank. The concentration of these residues was proportional to the length of exposure of the fuel to the diaphragm. Four preflight decisions were made to minimize this anticipated thruster problem: the fuel loaded in the SV was to be as clean as possible; at the time of lift-off the primary RCS tanks would be full of fuel; secondary RCS tanks would be empty to delay the onset of thruster degradation; vehicle activity would be minimized. Propellants would be loaded into the secondary RCS tanks and transferred to the secondary RCS only after the primary RCS started to degrade.[224]

The satellite vehicle for mission 1203 was mated to the booster vehicle on day R-13, and prelaunching checkout began and proceeded smoothly toward the planned launching date of 7 July 1972. Mission 1203 was successfully launched at 1046 PDT at the opening of the launching window, and the Titan-IIID injected the SV into the desired 96- by 137-nm orbit. As was the case with the previous flight, special experiments were conducted during the third launching and ascent to measure the contamination environment, which had caused over-specification battery temperatures on the first flight. Analysis of the data from these experiments confirmed that contamination occurred during the Titan-IIID SRM staging and was caused by the small thrusters at the front end of the launch vehicle that pushed the SV away from the core of the Titan-IIID at burnout.

The third HEXAGON mission, which was planned for 45 days of photographic operation followed by 15 days of solo operation, actually flew 69 days. Photographic operations were conducted on days 1 through 58, solo experiments and lifetime demonstration activities were conducted from day 58 to day 69, and the SV deorbited using Lifeboat-II during deboost. After ascent there was an anxious period when the lefthand solar array was very slow in erecting after release; however, it eventually deployed to the proper position.

During the first phase of mission 1203 (which was designated 1203-1 and connoted that exposed film which filled the first capsule [RV-1]), all camera operations showed normal characteristics, with no malfunctions experienced. RV-1 recovery during revolution 132 on day 9 was nominal; the capsule was air recovered, and a major section of the RV heat shield was retrieved from the water.

During segment 1203-2, "operational photography progressed normally until revolution 314 when there was an indication of minor disturbances in the aft camera's fine film path." Certain limitations in camera operations were established but "similar disturbances were reported on revolutions 348 and 350 but no further action was taken before recovery of RV-2 during revolution 359"[225] on day 22. Evaluation of the recovered film showed that, beginning on revolution 314, film in the aft camera had not tracked properly.

The first indication of an RCS thruster problem occurred during revolution 175, when a 100-degree temperature increase was observed on thruster 8. Despite this indication, thruster leakage was not significant until revolution 306. To dilute possible contaminants in the RCS, 50 pounds of propellant was transferred from the OAS tank to RCS tanks during revolution 331 without perceptibly affecting the leakage rate. Normal RV-2 separation, reentry, and recovery were carried out during revolution 359 on day 23.

During mission segment 1203-3, which lasted 14 days, the aft camera continued to show film-path disturbances. After a series of problems, aft camera operations were suspended on revolution 399 for the balance of the flight. This action was taken as a consequence of a film fold-over during revolution 364, which doubled the rate at which the take-up radius was increasing and, made a catastrophic failure probable. The overall quality of the film was reported to be "fair to good" with the aft camera performing noticeably better than the forward unit. Photographic quality improved in the aft camera and degraded in the forward camera as the mission progressed.

By revolution 385, the leakage of RCS-1 had increased to 0.25 pounds/revolution, and the thruster temperature had reached 1,000 degrees. By revolution 420, the RCS-1 userate was approximately 2.4 pounds/revolution (about 10 times nominal), and planning began for transferring fuel to RCS-2, which was accomplished during revolution 436. Mission segment 1203-3 was completed on day 36 during revolution 586 with normal reentry and aerial recovery of RV-3.

Mission segment 1203-4 lasted for 21 days with continued—but non-catastrophic—problems in the sensor and RCS. In this segment, the cameras operated normally to revolution 719, when the forward camera experienced a fold in the film similar to that suffered by the aft camera during segment 1203-3. Both cameras continued to operate; however, the aft camera was also used in a monoscopic mode to optimize film use and reduce camera risk.

In RCS-2, the propellant-use rates increased from 0.3 pounds/revolution during revolution 800 to 1.5 pounds/revolution during revolution 820. While it may not have been related, it was noted that "a pattern of increasing leakage was observed after revolution 801 when monoscopic camera operations, with one optical bar rotating, were interspersed with stereoscopic operations."[226] While monoscopic operations were part of the HEXAGON repertory, the reaction-control thrusters were used more frequently to compensate for the inertial imbalances involved. RV-4 reentry and aerial recovery occurred during revolution 924, ending mission segment 1203-4. On day 68, following a simulated mission late in the solo phase, the vehicle began tumbling. It was recaptured shortly thereafter and put under Lifeboat-II control. The SV reentered the atmosphere during revolution 1,104, successfully terminating the third HEXAGON mission. Other than the RCS and camera problems noted, all subsystems worked very well.

The fourth HEXAGON flight, mission 1204, was planned for a 60-day photographic phase followed by 15 days of solo operation. It actually flew 69 photography days followed by 22 days of solo experiments and lifetime demonstration activities. The Titan-IIID booster injected the satellite into a nominal orbit on 10 October 1972. The camera subsystem operated properly throughout the mission with some operational constraints to preclude the mistracking which had occurred during mission 1203. All film was recovered; it had an average ground resolution of 4.4 feet.[227] Because most photography was taken between ± 45-degree scan at a relatively low altitude, mission 1204 "provided the best overall image quality relative to previous HEXAGON missions."[228] All satellite vehicle performance was nominal except for the RCS and the attitude-control system (ACS). None of these anomalies affected mission success because of the availability and use of redundant equipment. The anomalies in the ACS included inertial-reference biases, a failure in a flight-control electronics assembly, and noise spikes.

The causes of the failures were identified and corrective action taken on subsequent flight hardware. Although satisfactory vehicle attitude and rate control was provided at all times during the 91-day flight, leaks in the primary system developed, as expected, and control was switched to the back-up system on day 26. No leaks were detected for the remainder of the flight. In the program evaluation, it was concluded that "elimination of the present RCS tankage, which has rubber diaphragms installed, stops the gross valve leakage problem experienced on previous flights."[229] On day 91, the SV was deorbited under ACS/RCS control during revolution 1,463.

On 20 January 1973, General Allen was transferred to the position of Chief of the Intelligence Community Staff, under DCI James R. Schlesinger. Allen was replaced as Director of SAFSP by Brig. Gen. David D. Bradburn, who had been Chief of the NRO Staff.

The fifth HEXAGON flight, mission 1205, began when the vehicle was placed into a nominal 85-by 158-nm orbit on 9 March 1973. Launched at 1300 PST near the close of the launching window, it had been delayed during countdown because of a multipathing problem between the satellite and the nearby Vandenberg Remote Tracking Station (RTS). For the first time, this SV carried the separate Mapping-Camera System (MCS) and its associated Mark-V reentry vehicle (RV-5). Itek had built the MCS and General Electric the Mark-V RV, which was quite similar to RVs used in the CORONA and GAMBIT programs. Mission planning was for a 70-day main-camera mission, including a concomitant 30-day mapping-camera mission, and five days of solo operation. About halfway through the mission, during 1205-3, a yaw-rate bias was observed, which persisted to some extent throughout the remainder of the flight. Panoramic camera velocity/altitude compensation capability was used to partially offset the yaw bias error. These compensatory adjustments were only partial, due to the relatively rapid changes in error. The yaw bias of 1.5 degrees caused a resolution loss of as much as 15 to 20 percent. All film in both cameras of the sensor system was used, and all four of the assigned RVs were recovered successfully. Even though the cameras performed very well, mission 1205 generally produced only fair image quality because of the excessive amount of haze and poor weather prevailing during the mission. The situation was compounded, to some extent, by the relatively late launching time and, therefore, post-noon acquisition times over targets.

The operation and performance of the first mapping camera were highly successful. Both the stellar and the terrain cameras functioned well, exposing 1,982 frames of film, with only minor anomalies, over a 42-day period. The resolution of the terrain camera was judged to be excellent throughout the mission. Evaluation of results indicated a quality level that significantly exceeded predicted values, based upon hardware acceptance test results. The evaluators[230] of the results rated "the image quality in ground resolution . . . outstanding for this scale. Numerous small manmade features were easily detected and occasionally identifiable; a baseball mound, small aircraft on taxiways, individual homes with driveways."[231] This was quite remarkable for a 12-inch focal-length lens at a 92-mile altitude. "The stellar photography provided adequate star images in both magnitude and quality,"[232] despite degradation by corona and solar radiation fogging. All RVs, including RV-5, performed properly and were air recovered without mishap. The SV was routinely deorbited over Shemya during revolution 1,139 on day 70.

The sixth HEXAGON satellite vehicle, mission 1206, was placed into a nominal 88- by 155.3-nm orbit on 13 July 1973. This was the second mission to carry the mapping-camera module and associated RV. Mission planning included a 45-day mapping-camera mission merged with a 75-day panoramic-camera mission, and five days of solo operation. On the fourth day of the mission, the SV experienced a primary-attitude control system (PACS) anomaly which caused a yaw bias. Control was switched to the redundant ACS (RACS) for the remainder of the mission. Except for 21 frames lost due to lack of a stellar-platen press operation, the mapping camera operated successfully throughout the mission with the film quality rated as very good. All the mapping camera film, including 61 frames of near-infrared (IR) film, was exposed and transported into RV-5, which was aerially recovered during revolution 683. The panoramic camera operated throughout the mission, and its RVs were aerially recovered on revolutions 310, 505, 926, and 1,202. The overall image quality of the panoramic camera system was rated as good. All the film was exposed and transported into the RVs, including 21,000 feet of SO-255 color film located in five separate segments on the aft camera and 500 feet of IR film on the forward camera. All solo tests were successfully completed and the satellite was deorbited during revolution 1,471 on day 92.[213]

On 25 August Colonel Krumpe was transferred to the Air Force Space and Missiles System Organization. He was replaced as HEXAGON program director by Colonel Ray E. Anderson, who had been serving as his deputy.

USAF Brig. Gen. David D. BRADBURN

USAF Col. Raymond E. ANDERSON

Section 11

HEXAGON Under New Management

Not long after the first successful HEXAGON flight, external circumstances made it necessary for DNRO John McLucas to consolidate all aspects of the program into Program A (SAFSP). The causal factor was the approval by President Nixon of program go-ahead for the ▓▓▓ electro-optical imaging program on 23 September 1971. ▓▓▓ had been selected for development as the next photoreconnaissance system; its planning, technology, and advocacy were by CIA's Office of Special Projects. Now that the program was cleared to proceed, its development management would require the concerted effort of the OSP/DDS&T staff. (The Air Force role in ▓▓▓ would be to provide launching and limited on-orbit support services.) Carl E. Duckett, now the DDS&T, agreed that he should offload work from OSP to ensure proper manning for ▓▓▓ management; as a result, all HEXAGON responsibility was transferred from Program B (OSP/DDS&T) to Program A (SAFSP). This rearrangement and its timing were directed in a message from the DNRO to the Director, CIA Reconnaissance Programs, and the Director, Program A.[234]

A principal transfer problem involved adjusting CIA/OSP contracts with the Perkin-Elmer Company. In this regard, the DNRO directed that the first buy of photographic payload systems (one through six) and the second buy (seven through 12) should remain the responsibility of CIA/OSP. The DNRO expected OSP to manage the contract for one through six (▓▓▓) to completion, but to "seek a convenient opportunity to transfer contract"[235] (▓▓▓) for payloads seven through 12 to Program A.

The plan for transferring CIA's HEXAGON responsibility to SAFSP was completed by ▓▓▓ (CIA/OSP), ▓▓▓ (CIA/OSP), and Col. Robert H. Krumpe (SAFSP) in March 1972.[236] Haas was deputy director of OSP; Patterson was the director, HEXAGON (SSPO), and Krumpe headed the HEXAGON SPO. The plan was concurred in by OSP director ▓▓▓ and approved by General Allen, Director, SAFSP (Program A). The first of July 1973 was set for completion.[237] It also envisaged that the execution of a tripartite agreement by the CIA, SAFSP, and Perkin-Elmer which would "substitute SAFSP for CIA/OSP as the customer, effective 1 July 1973"[238] for contracts ▓▓▓ (Flight Sensor Subsystems seven thru 12) and ▓▓▓ (Facilities Contract). The idea of a tripartite agreement had been recommended by ▓▓▓ chief, Contracts Staff/OSP, who had evaluated[239] the possible options. The terms of this agreement were carried out on schedule.

This transfer of responsibility was a complex and important operation, involving much more than contractual responsibility. It called for extensive communication to ensure that the new owner understood all matters ranging through engineering action, test and analysis methods, software management and support, and post-flight analysis, as well as issues of security responsibility and budget and fiscal actions. The record shows that representatives of both parties worked diligently and cooperatively to ensure that the program was neither hindered nor weakened by the transfer.

The effectiveness of the transition process was noted on 30 June 1973 in a message from General Bradburn, Director, SAFSP, to DNRO McLucas and to Leslie Dirks, Director of CIA Office of Development and Engineering (OD&E) that replaced OSP, in which Bradburn formally assumed "all responsibility for management of contract ▆▆▆▆▆▆ covering HEXAGON sensor subsystems seven through 12."[240] He extended his "personal thanks to the many people in OD&E who helped make this transfer in such an orderly and effective way." Bradburn noted that this milestone represented the final step in the transition and concluded by congratulating OD&E for the success of the program under its leadership, assuring the addressees that "we will do our very best to continue that proud record."

After the formal transfer date, CIA/OD&E continued to support SAFSP in both technical and business matters. In a typical case, because of the incentive nature of the Perkin-Elmer contract, OD&E researched its own records and gave SAFSP a complete rundown of fee penalty aspects involved in Perkin-Elmer delays on HEXAGON flights one thru six.[241]

The HEXAGON program continued to fly with ever-improving results. The seventh HEXAGON satellite (and the first Block-II panoramic camera and SBA) were placed into an 88- by 154-nm orbit on 10 November 1973. All ascent events were nominal and proper stabilization of the SV allowed deployment of the solar arrays at the first station contact. Preflight mission planning included a 45-day mapping-camera mission, a 90-day panoramic-camera mission, and a 30-day solo operation. The panoramic camera operated through the 103-day mission, and its RVs were aerially recovered on days 15, 38, 65, and 103. All the film was transported into the RVs, including 4,983 feet of SO-255 color film in RV-1 and 501 feet of FE-3916 infrared color film in RV-4. During the 1207-1 post-flight analysis, it was determined that in the panoramic camera a metering capstan resonance at peak Vx/h values was affecting image quality; in compensation, the perigee altitude was raised two miles on revolution 289. All other panoramic camera operations were normal. Mapping camera operations were also normal, and 98.4 percent of the film was transported to RV-5, which was recovered aerially on day 58. Solo tests were completed and the SV was deorbited on day 124, 13 March 1974.[242]

In addition to its normal mission, two new objectives were accomplished by mission 1207, both during segment 1207-4. First, ▆▆ second, a direct solar photography experiment was conducted to determine camera and film degradation.[243]

The remaining Block-II HEXAGON vehicles (missions 1208 through 1212) flew with remarkably few problems; the results are shown in the HEXAGON Operations summary at the end of this section. During the eighth flight, the attempt to aerially recover RV-1 was unsuccessful but the capsule was successfully retrieved from the water. Despite this problem—plus a few incidents with the panoramic camera system—all mission objectives were accomplished. Flight 1209 was normal, with the mission portion lasting a total of 129 days, followed by a 12-day solo phase and deorbit (on Lifeboat) on day 141. Flight 1210, after a few non-catastrophic problems in both the panoramic and mapping camera systems, flew for a total of 151 days.

Flight 1211 had panoramic camera problems; as a result, much of the 120-day mission was conducted in monoscopic photographic mode. During HEXAGON mission 1212, malfunctions in inputs from the solar arrays to the main battery bus of the satellite vehicle required power load management. During portions of the mission, operations of the mapping camera, doppler beacon, and ▓▓▓▓ were inhibited.

In August 1975, General Bradburn was transferred from Program A (SAFSP) to duty as the deputy commander, Electronic Systems Division, Air Force Systems Command. He was replaced by Brig. Gen. John E. Kulpa, who had been serving as director of the NRO Staff.

During the flights of the Block-II HEXAGON vehicles, work began on an improved version, known as Block-III (vehicles 13 through 18). Several areas in the spacecraft were changed. In the electrical distribution and power system, four Type-40 batteries replaced the seven batteries previously used (four Type-29, one Type-30, and two Type-31.[245] The new batteries were configured so that three would power the main bus; the other powered Lifeboat and could, if necessary, be switched to the main bus or to both. New thrusters, with extended lifetimes, were developed for the RCS. There was increased cross-strapping between the RCS and the attitude-control system. Two added tanks with ullage control were added to the orbit-adjust system (OAS), increasing the OAS propellant to 3,708 pounds.

The panoramic camera was provided with an improved emergency shutdown capability; now either camera could operate monoscopically, with both optical bars rotating and with revised film transport start-up equations and software.[246] The capacity of the nitrogen supply (supporting the airbars which served as film rollers in the film transport system and pressurized the film path) was doubled (from 34 to 68 pounds) and a "large looper" was added to decrease inter-operation film wastage, thus increasing the quantity of imaged film by about 20 percent.[247] Two film improvements were also made during Block-III. On SV-14, ultra-ultra-thin base (UUTB) film was flown instead of ultra-thin base (UTB), permitting even larger film loads to be carried. On SV-15, new mono-cubic-dispersed emulsion film was flown for the first time, significantly improving photographic performance.

To meet the Defense Mapping Agency's desire to use HEXAGON panoramic material to make maps, after the stellar-terrain camera was removed (mission 1216), Perkin-Elmer developed the solid-state stellar camera (S^3)—a system which utilized a light-sensitive charge-coupled device (CCD) at the focal plane in lieu of conventional film—to record the stellar field. S^3 flew on mission 1217 and through the remainder of the program.

The first Block-III HEXAGON vehicle, SV-13, set a new record for mission lifetime. Launched on 27 June 1977, it flew successfully for 180 days with only a few minor anomalies. It was deorbited on 23 December 1977, having successfully performed four times the original 45-day design goal.

The 14th HEXAGON mission, launched on 16 March 1978, also flew for 180 days. A malfunction of the aft camera of the panoramic payload on revolution 1,238 resulted in limited camera operation for the duration of the mission and, accordingly, the aft camera film supply was not depleted by mission's end.[248] Failure of the terrain camera's thermal shutter on revolution 869 resulted in an open thermal door for the remainder of the mission. The excessive power usage caused by these anomalies required power management, such as reducing the operation of ▓▓▓▓▓▓▓▓▓▓▓▓▓▓▓▓▓▓▓▓▓ and the redundant flight controls electronic assembly.

During the flight of HEXAGON mission 1214, Program Director Col. Ray Anderson retired from the Air Force. He was replaced, on 1 August 1978, by Col. Lester S. McChristian. During Anderson's tenure, the management of the HEXAGON and GAMBIT Programs had been combined within a single SPO. With the advent of longer but less frequent flights on both GAMBIT and HEXAGON, combining the two system program offices at SAFSP (and Aerospace Corporation) and eliminating the duplicate engineering staffs, test crews, and facilities at LMSC for these two programs saved substantial amounts of manpower and dollars.

USAF Brig. Gen. John E. KULPA

USAF Col. Lester S. McCHRISTIAN

SV-15 was launched on 16 March 1979. It flew for a record 191 days, 188 of which were "primary" (for photographic reconnaissance) and three were solo operations. It was deorbited on 22 September 1979. Although several anomalies were noted during the flight, none affected mission success. "Significant problems affecting the mission were the numerous remote tracking station failures and the failure of one of the extended command system programmable memory units (PMUs)." "The flight-support computers (CDC 3800s) experienced reliability and maintenance problems throughout the flight."[249]

Mission 1215 (SV-15) "was the first system to use type SO-315 film as the primary material. This was a newly developed Kodak fine-grain emulsion with UTR and possessing very-high-resolution characteristics. Both film supplies were successfully exposed and recovered; their imagery showed the best tri-bar resolution ever obtained with a HEXAGON system."[250] The mapping camera on SV-15 was operated for 119 days, the longest MCS mission to that date. As in the previous missions, thermal-door problems were encountered.

HEXAGON mission 1216 (SV-16) was launched on 18 June 1980. Following a successful mission, it was deorbited on 6 March 1981. During the (record) 161 days, the vehicle was placed in a parking orbit for 90 days (with only 72 camera operations) from late October 1980 through late January 1981—a period when winter weather

Launching of Mission 1216, 17 June 1980

limited broad coverage photography. This "storage" coincided with the end of mission segment 1216-3 and the beginning of segment 1216-4. During the active mission, both the panoramic camera system and the MCS performed well, except that the MCS terrain thermal door malfunctioned, as it had in two previous missions. This failure had no impact on the imagery, which was comparable to the good product of previous missions.[251] Late in the flight of mission 1216, the extended command system exhibited a series of failures which lent drama to the successful recovery of RV-4.

After its launching on 11 May 1982, the 17th HEXAGON vehicle, mission 1217—the fifth Block-III vehicle—performed well with 208 operational days (203 photographic, five solo). It was deorbited on 5 December 1982. SV-17, like the remaining vehicles, did not fly the mapping camera module; however, the panoramic camera system did include—for the first time—the system, which provided sufficiently precise vehicle-attitude information to permit panoramic photography to be used for some mapping, cartographic, and geodetic applications. In addition, "this was the first mission on which the sensor system used the large looper and modified film transport in order to reduce the amount of unexposed film between operations. As an example of film saving, wastage was reduced from 23 to 8 percent during the RV-1 (1217-1) segment, resulting in 10,400 feet of additional film for photography, compared to previous missions."[252]

The sensors performed well, with a few anomalies. One of consequence occurred on day 190 when "the A-side (forward camera) experienced an emergency shutdown (ESD) due to an apparent short. The result was loss of the A-side and subsequent monoscopic, single optical bar operations in the B-side for the remaining 13 days of the mission."[253]

Some problems were experienced in recovery. "The recovery of RV-1 was aerial. The recoveries of RVs 2, 3, and 4 had to be from the water, because of failure of the parachute-cone-bag cutters, which precluded the target cone from deploying." The failures were attributed to contamination in manufacturing: probably solder flux prevented the spring-loaded firing pin from driving a line-cutter, which held a cone-bag holding-line.[254]

On 19 January 1983, General Kulpa retired from active duty and was replaced as Director, Program A (SAFSP), by Brig. Gen. Ralph H. Jacobson. On 8 March 1983, Colonel McChristian was assigned as Jacobson's deputy (SP-2). McChristian was replaced as GAMBIT/HEXAGON Program Director by Col. Larry Cress. Cress was SPO Director for the remainder of the program, retiring from active duty on 22 May 1987.

HEXAGON vehicle 18, mission 1218, was launched on 20 June 1983 and demonstrated vehicle operation which "was generally excellent for all subsystems.""[255] It flew a 271-day (nine-month) primary mission plus five days solo and was deorbited on 21 March 1984. SV-18 carried 304,740 feet of film. The forward and aft cameras used 149,666 and 151,038 feet of film, respectively, during the 1,722 camera operations of the primary mission. These operations included 79 engineering tests.

USAF Brig. Gen. Ralph H.
JACOBSON

USAF Col. Larry
CRESS

Also 3,188 S^3 solid-state sensor[256] operations were conducted, including 21 S^3 activity detection (SCAD)[257] operations. The overall image quality ranged from very good to poor with the degraded portions attributed to haze, cloud cover, and veiling high cirrus. The sensor and S^3 systems showed no significant anomalies during the primary mission.

As an adjunct to the primary mission, SCAD tests were conducted on a non-interfering basis to demonstrate the feasibility of detecting nighttime activity using the S^3 subsystem. The flight-proven SSP software was modified to provide the vehicle maneuvers (pitch, roll, and yaw) required to point either the left or right S^3 boresight at the desired target. Algorithms were added to assure conformance to all system constraints, including maneuver time intervals, S^3 timelines, vehicle elevation and slant-range limitations, and solar vector angles relative to the sensor subsystem.

Results of the SCAD data evaluation:

a. Nighttime activity was detected. Thresholds were set, assuming clear weather, to obtain maximum intelligence without excessive data buffer overflow.

b. For some tests, particularly over Zone of Interior targets, a correlation could be made between known "ground truth" and the geometric distribution of light sources, as detected by the S^3 subsystem.

c. Tests conducted under full moon and snow conditions yielded usable data.

d. Refined attitude data were obtained by processing star "hits" imaged by the skyward-looking sensor.

e. Boresight pointing for a given target was repeatable to within one milliradian.

The four RVs were acquired with no damage to the capsules. The recovery of all RVs was aerial and normal with no recurrence of parachute-bag cutter problems, as in SV-17.

After the success of the Block-I, -II, and -III HEXAGON flights, it was disheartening to experience severe command system problems during mission 1219. Launched on 25 June 1984 and originally planned for a 302-day flight, 1219's duration was cut to 109 days. Three RVs were retrieved, containing 55 percent of the original film stock. It was necessary to deorbit the SV with the remaining RV-4, using Lifeboat, on day 109, 11 October 1984. SV-19 was the first Block-IV vehicle and the first with the Block-IV command system. The extended-command system portion of the command system contained plated wire memories in both of the parallel (PPMUs) which "directly related to the shortened mission as both PPMUs of the command system failed after numerous bit failures occurred during the flight."[258]

"These failures and subsequent safing of the vehicle and new operating procedures seriously reduced attainment of mission objectives. Uplink commanding and remaining hardware performed nominally."[259] After complete failure of the extended-command system, the minimum-command system was successfully used for all station contacts as well as the recovery and deorbit events.[260]

The HEXAGON flight program ended sadly on 18 April 1986, when the launching of the last vehicle (SV-20) was terminated by a catastrophic booster failure nine seconds after liftoff. A subsequent investigation by the Air Force Space Division (which was responsible for the Titan-34D) determined that a failure in the plumbing near a high-pressure pump in the "boat tail" part of the booster caused the explosion.

HEXAGON Operations Summary

Mission		Recoveries		Comments
1201				
Launched	6/15/71	RV-1	6/20/71	Chute problem, water recovery
Deorbited	8/06/71	RV-2	6/26/71	Aerial recovery
		RV-3	7/10/71	Chute failed, capsule lost
		RV-4	7/16/71	Aerial recovery
				50 percent film load
				2.7 feet resolution
1202				
Launched	1/20/72	RV-1	1/26/72	
Deorbited	2/28/72	RV-2	2/08/72	Forward camera failed
		RV-3	2/17/72	Monoscopic operation
		RV-4	2/28/72	Monoscopic operation
				2.7 feet resolution
1203				
Launched	7/07/72	RV-1	7/15/72	
Deorbited	9/13/72	RV-2	7/29/72	
		RV-3	8/12/72	Film path disturbances
		RV-4	9/12/72	Monoscopic operation
				RCS problems
1204				
Launched	10/10/72	RV-1	10/21/72	10,000 ft color film
Deorbited	1/08/73	RV-2	11/05/72	
		RV-3	11/23/72	
		RV-4	12/17/72	
1205				
Launched	3/09/73	RV-1	3/21/73	1st mapping camera
Deorbited	5/18/73	RV-2	4/05/73	Degraded photography due
		RV-3	4/19/73	to yaw error
		RV-4	5/11/73	
		RV-5	4/21/73	
1206				
Launched	7/13/73	RV-1	8/01/73	2nd mapping camera
Deorbited	10/12/73	RV-2	8/14/73	
		RV-3	9/08/73	
		RV-4	9/25/73	
		RV-5	9/24/73	
1207				
Launched	11/10/73	RV-1	11/24/73	All mission objectives
Deorbited	3/13/74	RV-2	12/17/73	satisfied
		RV-3	1/13/74	
		RV-4	2/20/74	
		RV-5	1/08/74	
1208				
Launched	4/10/74	RV-1	4/23/74	All mission objectives
Deorbited	7/28/74	RV-2	5/21/74	satisfied
		RV-3	6/26/74	
		RV-4	7/24/74	
		RV-5	6/09/74	
1209				
Launched	10/29/74	RV-1	11/17/74	All mission objectives
Deorbited	3/18/75	RV-2	12/23/74	satisfied
		RV-3	1/21/75	
		RV-4	3/07/75	
		RV-5	12/27/74	

HEXAGON Operations Summary (Continued)

Mission		Recoveries		Comments
1210		RV-1	6/24/75	Mapping camera
Launched	6/08/75	RV-2	7/29/75	power-relay failure
Deorbited	10/05/75	RV-3	9/04/75	limited performance.
		RV-4	10/06/75	Lifeboat used after
		RV-5	7/30/75	RCS degradation
1211		RV-1	1/07/76	Aft camera failure on
Launched	12/04/75	RV-2	1/27/76	day 20; monoscopic
Deorbited	4/01/76	RV-3	2/21/76	aft camera operations
		RV-4	3/29/76	resumed in 1211-4
		RV-5	2/02/76	
1212		RV-1	8/03/76	Power problem
Launched	7/08/76	RV-2	9/06/76	Limited terrain
Deorbited	12/13/76	RV-3	10/19/76	Sys. & ▓▓ during
		RV-4	12/09/76	early revolutions
		RV-5	9/08/76	
1213		RV-1	8/02/77	B-side shutter problem
Launched	6/27/77	RV-2	9/05/77	RV-2 water recovery
Deorbited	12/23/77	RV-3	11/04/77	All mission objectives
		RV-4	12/19/77	satisfied
		RV-5	10/17/77	
1214		RV-1	4/20/78	All mission objectives
Launched	3/16/78	RV-2	6/01/78	satisfied
Deorbited	9/11/78	RV-3	7/22/78	
		RV-4	9/09/78	
		RV-5	7/11/88	
1215		RV-1	4/26/79	PMU-5 hardware failure
Launched	3/16/79	RV-2	6/16/79	RV-3 drogue mortar end cap
Deorbited	9/22/79	RV-3	7/31/79	penetrated capsule
		RV-4	9/19/79	
		RV-5	7/12/79	ST thermal door failure
1216		RV-1	7/24/80	ST thermal door failure
Launched	6/18/80	RV-2	9/07/80	PMU-B failure
Deorbited	3/06/81	RV-3	10/24/80	Insensitive SGLS-1 receiver
		RV-4	3/05/81	
		RV-5	10/14/80	RDA failure
1217		RV-1	6/15/82	Forward Camera (A) failure on day
Launched	5/11/82	RV-2	8/02/82	190; Mono-B for remainder of
Deorbited	12/05/82	RV-3	9/29/82	mission
		RV-4	11/30/82	Water recoveries of
				RVs 2, 3 & 4
				1st S^3
1218		RV-1	8/24/83	Bit flip in Hybrid
Launched	6/20/83	RV-2	11/07/83	PPMU-A program section
Deorbited	3/21/84	RV-3	1/09/84	S^3
		RV-4	3/16/84	
1219		RV-1	8/05/84	RV-1 water recovery
Launched	6/25/84	RV-2	9/24/84	Serious failures in PPMU
Deorbited	10/11/84	RV-3	10/11/84	-A, B caused early
		RV-4	not used	mission termination
				S^3
1220				Entire mission lost due to failure of
Launched	4/18/86			Titan booster 9 seconds after liftoff

THE HEXAGON STORY

HEXAGON — A Unique Intelligence Asset

During its 13-year life, HEXAGON provided a unique collection capability which may never again be achieved by US imagery satellites. Its ability to cover thousands of square nautical miles with contiguous, cloud-free, high-resolution imagery in a single operation* provided US intelligence users and mapping, charting, and geodesy (MC&G) organizations with vast amounts of nearly simultaneous contiguous coverage. Order-of-battle information across entire Soviet military districts could be achieved in a short timeframe. Sino-Soviet military tactics could be studied and determined by analyzing imagery of Warsaw Pact, Soviet, and Chinese large-scale exercises. HEXAGON provided the best MC&G support ever furnished to the user community—large-scale contiguous imagery within specified geometric accuracies.

*The term "single operation" refers to one "camera on - camera off" cycle. These cycles varied considerably in length of operation.

HEXAGON Coverage Achievements

KH-9 Missions			Coverage Accomplishments (Million square nm)	
Mission Number	Launch Recover Dates	Life time Days*	Worldwide	Communist Countries and Missile East
1201	15 Jun 71 - 16 Jul 71	31‡		
1202	20 Jan 72 - 28 Feb 72	39		
1203	7 Jul 72 - 12 Sep 72	57		
1204	10 Oct 72 - 17 Dec 72	68		
1205	9 Mar 73 - 11 May 73	63		
1206	13 Jul 73 - 25 Sep 73	74		
1207	10 Nov 73 - 20 Feb 74	102		
1208	10 Apr 74 - 24 Jul 74	105		
1209	29 Oct 74 - 7 Mar 75	129		
1210	8 Jun 75 - 6 Oct 75	120		
1211	4 Dec 75 - 29 Mar 76	116		
1212	8 Jul 76 - 9 Dec 76	154		

*Imaging days on orbit, not counting days of launch as an actual day on orbit.
†COMIREX target population has ranged from about ▓ in the earlier missions to about ▓ on the most recent missions.
‡RV-3 was lost on 1201.

Mapping Camera (12-inch Terrain) Coverage	
Mission	(Thousands of Square Nautical Miles)
1205	
1206	
1207	
1208	
1209	
1210	
1211	
1212	
1213	
1214	
1215	
1216	

One of HEXAGON's most significant contributions to the US security posture was the confidence it provided national leaders in negotiating arms-limitation agreements with the Soviets and conducting continuing negotiations for future treaties. HEXAGON was of paramount importance in confirming or denying Soviet strategic weapons development and deployment. Any new Soviet ICBM complex or development—such as mobile missile deployment—could be detected quickly. New construction of antiballistic missile—(ABM-) related facilities or production of nuclear submarines capable of ballistic missile launchings was monitored closely. Inactivation of outdated weapons systems could be observed. This information was invaluable at the international negotiating table.

In January 1977, the ▓▓▓▓ electro-optical imaging (EOI) system came on line. Although primarily planned as a high-agility, high-resolution system with near-real-time (NRT) capability, this system could also provide broad-area coverage similar to HEXAGON. It, therefore, was considered to be the replacement for HEXAGON—a system conceived and developed under rather difficult circumstances—but one which performed well above and beyond user expectations.

Section 12

HEXAGON Financial Summary

The total cost of the 20-flight HEXAGON program, including the CIA-managed portion, beginning with FY66 and ending with FY86, was $3,262,000,000.[261] Of this, the CIA funding totaled ▓▓▓▓ which largely went for the development and production of the first 12 sensor units at Perkin-Elmer ▓▓▓▓. The CIA figure included, in addition to Perkin-Elmer payload costs, the amount of ▓▓▓▓ for special facilities and test equipment for Perkin-Elmer, ▓▓▓▓ for the SSPO SETS contractor (TRW), ▓▓▓▓ for field support. Of the total $3.26 billion cost, approximately ▓▓▓▓ was spent as DoD Secret or "white" funds; the balance, ▓▓▓▓, was spent as "black" funds. This translates to an average cost per flight of ▓▓▓▓ and an average cost per photographic day[262] of ▓▓▓▓.

Using available data on number of cloud-free unique targets taken by some of the missions, as well as the number of cloud-free square nautical miles covered on those missions, and assuming these are representative, the cost per unique cloud-free target[263] was ▓▓▓▓ and the cost per unique cloud-free square nautical mile[264] was ▓▓▓▓.

Of the major contractors, Perkin-Elmer received ▓▓▓▓ for photographic payloads and related items, Lockheed Missiles and Space Company, Inc., received ▓▓▓▓ for the SBA and related goods and services, Itek was paid ▓▓▓▓ for the mapping-camera subsystem, and GE ▓▓▓▓ for the command subsystem. The cost of launching the spacecraft totaled ▓▓▓▓, of which Martin-Marietta Company received ▓▓ for Titan-III hardware and launching services. As the technical overseer of the program, the Aerospace Corporation was paid ▓▓ million.

Section 13

A Goodly Heritage

The history of satellite reconnaissance began with a visionary RAND study, sponsored by a newly designated military service, the Air Force. The vision could not become reality until a means was found for boosting heavy loads into earth orbit. This means finally appeared in 1958 in the form of IRBM-class boosters, and a concerned US President immediately seized the opportunity to sponsor this new possible method for observing, from space, activities in hostile, denied areas.

A Growing Technical Excellence

Only 30 months after Eisenhower's decision, the CORONA satellite made its first successful flight, delivering photography at resolutions of 30 to 40 feet. With improved camera and film, CORONA resolution soon moved to 10 feet; finally, ground resolutions of 6 to 10 feet became common, with area coverages of over 8,000,000 square nautical miles.

CORONA was a search system, designed to answer the question, "Is there something there?" The Intelligence Community always has a follow-on need, categorized as surveillance, which says, "There *is* something there! We now want to watch it, learn more about it, and, if possible, identify and classify it." Once again, Eisenhower took leadership in sponsoring a new imaging satellite system, and within three years, the surveillance "bird" was producing pictures. Early flights of this GAMBIT system delivered photography at resolutions of two to three feet; eventually, these numbers improved to ▓▓▓▓▓▓ Finally, it became routine to expect GAMBIT to cover ▓▓▓▓▓ targets; when mission-life was extended to four months in the closing days of the GAMBIT-3 program, almost ▓▓▓ targets would be covered.

Six years after the CORONA decision, it was reasonable to envision a follow-on reconnaissance system which would combine the capabilities of CORONA (search) and GAMBIT (surveillance). This volume of the NRO history has recounted the new development that produced that system (HEXAGON) and detailed its impressive performance as an intelligence collector.

A Growing International Acceptance

Satellite reconnaissance began operation without benefit of a judicial code—such as the well-defined international law of the high seas—for establishing the legitimacy of such activities in space. In the 1960s, there was always a question as to whether the Kremlin would object to an operational reconnaissance satellite. As a corollary, if the Kremlin *did* object, would its reaction culminate in actual interdiction?

These and other concerns were shared by a number of DoD and State Department officials including DNRO Joseph Charyk who, in discussions with Deputy Defense Secretary Roswell Gilpatric and Under Secretary of State U. Alexis Johnson, urged the formulation of a national policy on satellite reconnaissance. The primary objective of such a policy would be to avoid, blunt, or at least defer confrontation with the Soviet Union. It was agreed that "we must avoid provoking 'them' into such objection." It was an accepted fact that the problem would be just as serious whether the "them" turned out to be (unexpectedly) a friendly country or (more expectedly) a Communist puppet nation. But the main case—a properly planned response to strenuous objection by the USSR—should be the primary consideration of US policymakers.

The initial step taken by the DoD to control information to news media on all military space flights—actual and proposed—was embodied in the "Gilpatric Directive" (DoD 5200.13 of 23 March 1962). This action placed a security blanket over all details of all military space programs and, in consequence, severely limited release of information regarding these activities. Gilpatric subsequently sent a proposed paper on "National Policy on Satellite Reconnaissance" to President Kennedy's Special Assistant, Gen. Maxwell Taylor, recommending that the subject be given immediate consideration. In response, the NSC issued National Security Action Memorandum (NSAM) 156, which set up a committee under the chairmanship of U. Alexis Johnson to develop US policy with respect to US reconnaissance programs and outer space. Among other things the policy aimed to maintain unilateral freedom of action to conduct space operations and to prevent foreign political and physical interference with the conduct of these operations.

The report of the NSAM 156 Committee and its recommendations for US policy on outer space were discussed at the 10 July 1962 meeting of the NSC, which approved 18 points of policy.[265]

Additionally, the BYEMAN and TALENT-KEYHOLE security systems—put in place specifically to protect all aspects of reconnaissance operations and products—were deemed to be still other important factors in keeping the NRO program obscure and inoffensive to the international community.

The most effective protective measure of all was furnished by the Soviets themselves on 12 May 1962, when they launched their own reconnaissance satellite, Cosmos, under similar close security. The existence of this spacecraft in orbit symbolized tacit acceptance of "freedom of space;" in Washington one could imagine echoes of Eisenhower's 1955 "Open Skies" plan.

The final symbol of acceptance occurred a few years later, when both the United States and Soviet Union adopted a soothing euphemism for reconnaissance satellites: "National Technical Means of Verification" (NTMV).

A Founder's Accolade

Col. (later Lt. Gen.) Andrew J. Goodpaster was Staff Secretary to Eisenhower during the bulk of that President's administration—1954 to 1961. He joined the President in conference with nearly every visitor, sitting unobtrusively at the side, jotting an occasional note. At the end of the conference, he would accompany the visitor to an anteroom and review key points and decisions made by the President; then his handwritten notes would go into a special file box for ready, definitive reference.

Goodpaster was well aware of Eisenhower's concern over surprise nuclear attack. He had observed—and perhaps participated in—the President's early decision that no task "transcended in importance that of trying to devise practical and acceptable means to lighten the burden of armaments and to lessen the likelihood of war."[266] He had attended White House conferences leading to the building of the U-2; later it had been his sad task to advise the President of Gary Power's disaster. He had also recorded Eisenhower's decision to build CORONA and is regarded, in that program's folklore, as a patron and founder.

One afternoon years later, in the summer of 1964, a request went to the office of the DNRO to provide some "satellite information" to the Assistant to the Chairman, Joint Chiefs of Staff, Maj. Gen. Andrew J. Goodpaster. The NRO Staff's Deputy for Plans—a graying Colonel—was sent in immediate response and was greeted cordially and disarmingly by Goodpaster with a paternal "Come right on in, son!"

Goodpaster's questions were brief, direct, and sequential; he was still the ultimate staff officer. What could CORONA do? Was CORONA vulnerable? Did it have potential for improvement? Was the program adequately funded? In a few minutes the brisk interrogation came to an end. Goodpaster paused briefly, in thought. Then, in a softer tone, he said, "Tell your people that they have done a mighty work—well beyond what we ever dreamed was possible. Keep *on* moving ahead; always ahead. You know, your group is so secret that it will never hear any public praise. I think it may be enough for you to know that you've put us in a position to keep watch on the Bear. I have the belief that you have given us hope for a quarter century of peace with that Bear."

As these lines are written 24 years later Goodpaster's quiet assessment, so visionary in 1964, is very close to coming true.

Appendix A

HEXAGON and the Intelligence Community

National Intelligence Requirements Management

The first HEXAGON was launched on 15 June 1971. Its function was to fulfill overhead imagery requirements developed by the Intelligence Community's Committee on Imagery Requirements and Exploitation (COMIREX).

COMIREX had been established on 1 July 1967,[267] with these functions:

> In accordance with policies approved by the United States Intelligence Board (USIB), the Committee shall advise, assist, and generally act for the USIB on matters involving the coordinated development of intelligence guidance for imagery collection by overhead reconnaissance of denied areas and, as set forth in the National Tasking Plan (NTP) for the Exploitation of Multi-Sensor Imagery, on matters involving the exploitation of imagery.

COMIREX was a follow-on to the Committee on Overhead Reconnaissance (COMOR), which had been established in 1960 to manage overhead reconnaissance intelligence requirements. The primary change between the committees was an expansion of COMIREX's roles and mission in the imagery arena and the assignment of COMOR's SIGINT responsibilities to a new USIB unit, the SIGINT Overhead Reconnaissance Subcommittee (SORS).

The membership of COMIREX was comprised of designated officials of the departments and agencies that constituted the Intelligence Community and were represented on the USIB: CIA, DIA, NSA, State, Army, Navy, Air Force, Defense Mapping Agency, and Atomic Energy Commission, now part of the Department of Energy. Consultants were appointed from agencies that were doing systems development and imagery exploitation: the National Reconnaissance Office (NRO) and the National Photographic Interpretation Center (NPIC). (See Graphic 1.)

In 1975, the Civil Applications Committee (CAC) was established with representation from the Departments of Commerce, Interior, Agriculture; the Environmental Protection Agency (EPA); and the Agency for International Development (AID) to apply satellite imagery to civil requirements. An earlier informal group, known as ARGO, had operated on an ad hoc basis since 1966. COMIREX was charged with overseeing activities of the CAC and ensuring that national imagery security policies were followed in the use of any authorized imagery. Only domestic imagery was eligible for use by CAC agencies, except for AID. Imagery of national disasters, such as drought, famine, and floods, was provided to assist the US Government in determining humanitarian aid requirements. HEXAGON's broad area coverage capability was ideally suited to satisfying disaster coverage needs such as floods and earthquakes, and also civil mapping requirements; it, therefore, was more frequently used than any other overhead system to satisfy CAC requirements.

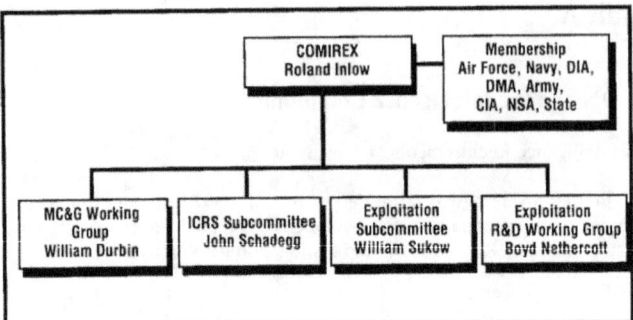

Graphic 1. Committee on Imagery Requirements and Exploitation

The day-to-day management of the Intelligence Community's collection and exploitation requirements was handled by two COMIREX subcommittees: the Imagery Collection Requirements Subcommittee (ICRS), responsible for managing collection requirements, and the Exploitation Subcommittee (EXSUBCOM), responsible for providing exploitation guidance to national exploitation centers.

By 1971, the COMIREX requirements interface with the system operator—the NRO—was through the NRO's Satellite Operations Center (SOC) in the Pentagon. The NRO developed a HEXAGON Reports Control Manual (RCM) that specified at what time in the mission cycle—both pre- and post- launching—COMIREX would furnish requirements data. For example, the desired film load for a specific mission had to be furnished to the NRO not later than launching minus ▮ days and the initial mission requirements at launching minus ▮ days. These examples indicate the extensive pre-mission planning phase of each mission. With a film load in excess of 200,000 feet and consisting of four or more different film types with different film thicknesses distributed throughout the film supply, the production, splicing, and stacking of film became a major technical undertaking. Similarly, the pre-mission planning phase for HEXAGON was far more extensive than that for GAMBIT. Numerous mission simulations and iterative reviews with ICRS were required to arrive at optimal projected requirements for each mission. After launching, the RCM specified timelines for real-time activities such as weather forecasts, "bucket" recovery schedules, and film deliveries to processor.

Flight Operations Management

HEXAGON flight operations were managed by the Secretary of the Air Force Special Projects Office ▓▓▓ at the Satellite Test Center in Sunnyvale, California. There the targeting software (T'Unity) was run and the actual camera operations selected; then the vehicle and payload commands were generated and transmitted to the orbiting HEXAGON vehicle through the Satellite Control Facility's worldwide tracking and control network. All this was done in the minimum time required in order to take advantage of the best possible weather forecast data. Until 1977, the SOC acted as the interface between the requirements manager, COMIREX, and the systems operator, ▓▓▓ of the NRO.

In 1977, all SOC responsibilities for HEXAGON operations were transferred to ▓▓▓ at the Satellite Test Center in Sunnyvale, California. This improved the effectiveness of HEXAGON operations by creating a more efficient, direct interface between the requirements manager (COMIREX) and the system operator ▓▓▓. Also, by eliminating an NRO middleman, some timelines were shortened and the possibility of misinterpreting requirements (especially those which were special or ad hoc) was lessened. Another influencing factor was the installation of a COMIREX Automated Management System (CAMS), discussed later, within the ▓▓▓ facility; CAMS provided a direct tasking link between COMIREX and ▓▓▓. In recognition of its broad responsibilities and authorities, ▓▓▓ was, in 1981, designated an Operating Division (OD-4) under the Secretary of the Air Force Special Projects (SAFSP) Office in Los Angeles.

OD-4 played a key role in the success of the HEXAGON program. A primary factor was OD-4's application of the human judgment element to the computer-generated mission plan and on-orbit targeting. This was important for HEXAGON operations, even more so than in other programs, because film management played such a prime role in each mission's success. The huge amount of film carried by HEXAGON was in danger of being quickly and inefficiently expended, if not managed carefully. Missions of 260 days duration were achieved against a design specification of 45 days. As mission durations extended, the number of missions could be reduced; thus sound operational management of mission resources became critically important.

Pre-mission planning was more important to HEXAGON's success than it had been to other programs because it established the level of film allocation by requirement types, requirement priorities, weather thresholds, and operational strategies, each of which would have a major effect on the course of the mission. Extensive iterations between the NRO and ICRS were accomplished to ensure an optimal mission plan for presentation to COMIREX for final approval. At the completion of pre-mission planning, the Intelligence Community had high confidence in the accuracy of projected levels of satisfaction against standing, special, and mapping, charting, and geodesy (MC&G) requirements. (Standing requirements defined collection objectives that took into account ongoing scheduled intelligence needs in a form consistent with the capabilities of existing or programmed systems. Special requirements provided day-to-day adjustments to collection tasking through Community mechanisms to reflect and respond to immediate or changing intelligence needs.)

Application of the human judgment factor paid off most significantly in on-orbit operations. Although T'Unity targeting software was used to provide recommended targeting operations, each T'Unity selection was manually reviewed. Consideration was given to actual predicted weather, satisfying mission requirements as a whole, climatology for future accesses, and desired mission duration. These subjective factors could not be programmed into the targeting software but were highly important in the final determination of camera operations; in fact, most software-generated targeting selections were modified after review by the OD-4 operations team.

HEXAGON Imagery Security Policy

HEXAGON imagery and imagery-derived products were controlled within the TALENT-KEYHOLE (TK) security system. Access to TK-protected information required a special security clearance and an authenticated need to know. In the early 1970s, the increased utility of satellite-derived information made it essential to provide more of it to lower-echelon military and other Intelligence Community users outside the TK compartment. Accordingly, in November 1973, President Richard Nixon approved recommendations by DCI William Colby that modified some of the strict security controls on the imagery satellite program. Specifically, the DCI was authorized to remove from TK controls, after consultation with the Secretary of Defense, such photographic products as he deemed appropriate, provided that the products removed were appropriately classified and did not reveal the sensitive technical capabilities of current or future intelligence satellite programs. As a result of this authorization, most of the product—except original-format film—and almost all of the information derived from it became available to US intelligence users at the Secret level outside the TK security control system. This action significantly increased the use of intelligence from the HEXAGON program.

The chairman of COMIREX managed the TK security system for the DCI. A basic policy objective for HEXAGON product was increased usage outside the TK security control system in meeting requirements of the Intelligence Community, the military forces of the United States and its allies, and the federal mapping agencies. The BYEMAN control system, which handles access to operational and programmatic data on NRP programs, is managed by the NRO and was unaffected by modifications to the TK security system.

System Requirements for the HEXAGON Photographic Reconnaissance System

HEXAGON was the first overhead reconnaissance system with development and system characteristics defined primarily by national intelligence requirements. On 21 June 1966, COMOR forwarded to the USIB[268] specific requirements for a new search/surveillance system to replace CORONA. The stated requirements were:

"a. **Resolution and Swath.** The requirement for a capability for search with a continuous stereoscopic swath width at least equivalent to KH-4[269] (150 to 180 nm) and a resolution equivalent to KH-7 (3 to 5 feet over the total format), as approved by the USIB on 31 July 1964, is reaffirmed.

b. **Obliquity and Stereo Convergence.** KH-9 should be designed to provide photography from vertical to between 45 degrees and 60 degrees obliquity. The stereo convergence angle should be no less than 20 degrees and no greater than 45.

c. **Search Mission.** KH-9 should have the capability to provide stereoscopic, cloud-free (about 90 percent) photography of about 80 to 90 percent of the built-up areas of the Sino-Soviet block (approximately 6.8 million square nm) semiannually and should provide similar coverage of about 75 percent of the undeveloped areas (2.8 million square nm) annually. It should be noted that this requirement differs from that approved by USIB on 19 March 1965 (USIB-D-41.14/229; COMOR-D-13/43) and that it is based on the results obtained and general satisfaction with search coverage acquired over the last 18 months with the KH-4. In addition to search of the Sino-Soviet bloc, KH-9 should provide the capability to acquire coverage of contingency areas in other parts of the world on demand.

Present areas requiring this coverage are Indonesia, the Middle East, Southeast Asia, and parts of North Africa. We do not expect this requirement to exceed 3 million square miles per year.

d. **Surveillance Mission.** In recognition of the capability of KH-9 to obtain high-resolution area coverage when meeting the specifications [in item a] above, we believe it appropriate to specify frequency of coverage in terms of surveillance of geographic areas representing target clusters rather than in terms of surveillance of individual point targets. Based on target distribution, we have identified about ▮ clusters ranging in size up to ▮▮▮▮▮ areas in which approximately ▮ percent of current targets are located. As new targets are added to the list, it is expected that the great majority will also fall in these same clusters. Although the bulk of these areas are located within the Sino-Soviet bloc, several of similar size fall outside this area. These target clusters, each of which contains a variety of target category types, should be considered dynamic and therefore subject to change as experience with KH-9 is acquired. For planning purposes, however, we believe that surveillance of about ▮ percent of these areas quarterly should be accomplished, especially since the KH-8 high-resolution spotting system can be employed to round out coverage or to obtain additional coverage as may be deemed necessary.

e. **Flexibility**. As presently described by the NRO, KH-9 will provide missions of 30 to 50 days duration. This long life, while providing the capability to acquire greater amounts of cloud-free photography through taking advantage of weather opportunities, includes the liability that the information derived will be old when received unless provisions are made to recover, process, and read out missions in increments. We believe that each recovery vehicle increment should contain no more than 10 days' coverage[270] and that there should be additional flexibility provided to recover portions of a mission in less than 10 days on demand even at some sacrifice in total mission coverage.

In order to avoid acquiring a great quantity of coverage in a few days and then being faced with a long period with no search/surveillance being conducted, we believe a capability such as a ▮▮ Were these capabilities available, readout of photography acquired early in a mission could be used to influence collection later on in the mission, the system would have the capability to respond to special events or to current intelligence needs, and excessive peaks and valleys in the rate of collection could be avoided.

f. **Standby Capability**. In order to assure that search/surveillance is conducted without undue time delays in coverage, standby vehicles at about R-3[271] days should be available to provide backup for possible failures and to provide emergency contingency coverage during times when no vehicles are on orbit.

g. **Mapping and Charting**. For KH-9 photography to be used directly in the preparation of maps and charts, it must contain the strong geometry required to meet the horizontal and vertical accuracy for large- and medium-scale maps and charts of which the most demanding is the large-scale (1:50,000) topographic map. These maps require a relative horizontal accuracy of 85 feet and a vertical accuracy of 16 to 33 feet over a distance of 10 to 20 miles. An accurate photogrammetric control network extending 500 miles in any direction within specified regions is essential for the development of an orderly production of coordinated series maps and charts. KH-9, in addition to providing search/surveillance as stated [in items c and d] above should also provide coverage of about 7 to 10 million square miles of the free world each year. This requirement usually can be satisfied by one-time coverage supplemented by re-coverage of relatively small areas (see COMOR-D-13/65 for additional statement of requirements)."

COMOR recommended that USIB approve the stated requirements and forward them to the NRO for use in system design. It was also requested that the NRO provide COMOR with information on those specific requirements that could be exceeded appreciably at negligible increased cost and/or those specific requirements which, if reduced, would result in substantial program savings or in substantial improvements in other requirements areas. USIB approved the COMOR recommended requirements on 20 July 1966.[272]

There were several other key HEXAGON development decisions that were responsive to USIB/COMOR identified needs during the development phase. These primarily were related to improvements in HEXAGON's capability to meet MC&G needs and included the addition of a 12-inch focal length stellar-index (SI) camera to the system at an estimated cost of ▓▓▓▓. The SI camera was needed to meet the Defense Mapping Agency's stated photogrammetric control network requirements, established as:[273]

- horizontal error of 40 feet over 20 miles and 400 feet over 500 miles
- vertical error of 10 to 20 feet over 20 miles and 80 feet over distances up to 100 miles

A Doppler beacon and accelerometer were also added to the system to support MC&G requirements and ensure the required horizontal and vertical accuracies were met. The NRO estimated the additional development cost at ▓▓▓▓, plus ▓▓▓▓ per mission. The stellar-terrain camera system and Doppler beacon were added starting with mission 1205 in March 1973.

The NRO met or exceeded COMOR's requirements, as shown in this Table.

Category	Stated Requirements	Achieved Capability
Resolution	3-5 feet	2-3 feet
Swath	150-180 nm	300 nm
Obliquity Stereo	45° – 60° obliquity	60° obliquity
Convergence	20° – 45° convergence	20° convergence
Search Mission	Built-up areas (6.8 million sq nm) Coverage 80-90% semiannually	Usually achieved*
	Undeveloped areas (2-8 mm sq nm) Coverage 75% annually	Usually achieved*
	Contingency areas (3.0 million sq nm) annually	Exceeded
Surveillance Mission	Target clusters - coverage 80% quarterly	Usually achieved*
Flexibility	30-50 days mission duration Recovery on demand	30-270 days Achieved
Standby Capability	R-3 day standby capability	Requirement deleted due to cost considerations
Mapping and Charting Geometry	• 85 ft horizontal accuracy	Achieved
	• 16-33 ft vertical accuracy over 10-20 miles	Achieved
	• 7-10 million sq nm free world coverage annually	Exceeded

*The frequency of missions flown determined the level of satisfaction. The originally planned launching schedule, if maintained, would have consistently met these requirements.

COMIREX Automated Management System (CAMS)

As the definition of intelligence requirements grew more complex and HEXAGON and other NRO satellite programs delivered increasing amounts of imagery, the need for an automated, interactive requirements management system became mandatory. Although some form of computer support had been available to the Community from the earliest days of the CORONA program, all such support was in the form of offline programs that were useful in mission planning and requirements analysis but had little utility for near- real-time management of requirements during the course of a mission. Furthermore, the Community members could not directly access the national data base to retrieve information on requirements, imaging attempts, past coverage, and so forth. The shortfall was eliminated in 1976, when CAMS became operational. For the first time, Intelligence Community members could, from a CAMS computer terminal located in their own facility, nominate a collection or exploitation requirement. If the requirement was of a time- sensitive nature, such as a ▓▓▓▓▓▓▓▓▓▓▓▓▓▓▓▓▓▓▓▓▓▓▓▓▓▓▓▓▓ ▓▓▓▓, the COMIREX staff could react immediately by tasking the NRO to attempt coverage of the border area on a priority basis. Provided that an imagery satellite was on orbit, it could be tasked against such a requirement ▓▓▓▓▓▓▓▓▓▓ than, as previously, hours or days. (The CAMS network and environment are depicted in Graphics 2 and 3.)

National Imagery Exploitation Responsibilities

In January 1961, National Security Council Intelligence Directive (NSCID) Number 8 established responsibility and procedures for the conduct of imagery exploitation in response to national foreign intelligence needs. The directive created a National Photographic Interpretation Center (NPIC) for priority exploitation of satellite imagery and charged the Center with providing common imagery support services to imagery exploitation organizations within the Washington, DC, area. NPIC was also charged with maintaining an up-to-date, consolidated file on imagery-derived target data to serve national and departmental needs. The NSCID provided that imagery exploitation requirements that were uniquely departmental in nature, for example DoD studies, were not the direct responsibility of NPIC, but were to be undertaken by the departments concerned. Those agencies without photointerpretation capabilities, for example State Department, could call upon NPIC to meet their needs.

Consistent with NSCID Number 8, an NTP for the Exploitation of Multi- Sensor Imagery was issued in January 1967. This plan defined the specific roles and responsibilities of Intelligence Community imagery exploitation organizations—NPIC, CIA, DIA, Army, Navy, and Air Force—in response to national imagery exploitation requirements. National requirements for imagery exploitation by the Intelligence Community were to be developed and managed by COMIREX.

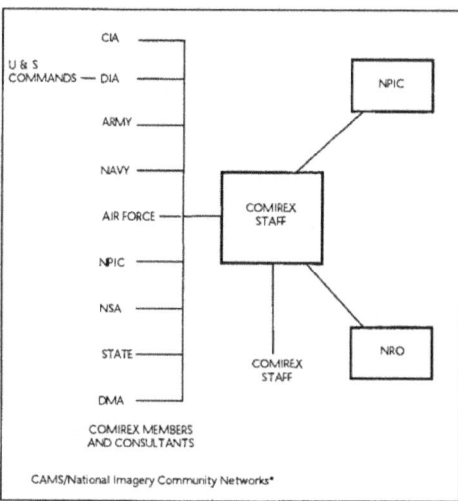

Graphic 2. CAMS National Imagery Committee Network

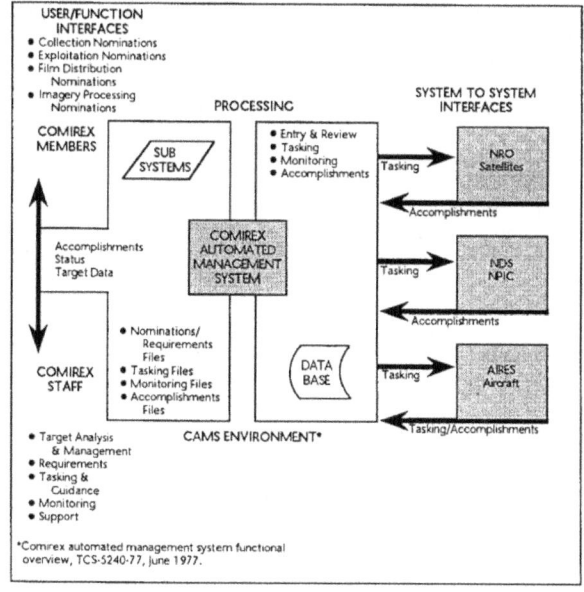

Graphic 3. CAMS Environment

Film Dissemination Responsibilities

Requirements for disseminating HEXAGON imagery were prescribed by the EXSUBCOM of COMIREX in response to Community needs. Imagery products included film, exploitation data, and printed matter. Additional imagery-related material included data on target coverage, film indexing, camera performance evaluation, mapping, cloud coverage/general weather, requirements satisfaction, and overall system performance evaluation. This process was dynamic, continuously supplying data, whether it was film products, information on operational control and management of a mission underway, future mission planning data, or exploitation end products.

National Photographic Interpretation Center (NPIC)

NPIC played a primary role in the success of overhead imagery programs. Collecting large volumes of HEXAGON imagery would serve little purpose without a dedicated and responsive organization to exploit and report on both the key intelligence information derived from each mission and routine information, such as order of battle, on which continuing and long-range intelligence decisions could be based. NPIC's search of and reporting on the Soviet Union following each mission was a key input for US Strategic Arms Limitation Talks (SALT) monitoring and ongoing SALT negotiations. During times of international crises, such as Middle East hostilities, India–Pakistan border tensions, and so forth, a HEXAGON mission would image the area of concern, and NPIC photointerpreters would be sent to Eastman Kodak to conduct immediate readout of the area of interest in order to provide national policymakers and the Intelligence Community the most current information available. On occasion a mission "bucket" might be returned earlier than planned or extended on orbit to satisfy urgent current intelligence needs. NPIC provided outstanding readout in satisfying national intelligence requirements throughout the HEXAGON program.

It is appropriate to make special mention of the first Director of NPIC, Mr. Arthur C. Lundahl. A superb technician in photographic interpretation and photogrammetry, Lundahl used the talents of individuals from such diverse disciplines as photointerpretation, photogrammetry, printing and photo processing, automatic data processing, communication and graphic arts, collateral and analytical research, and technical analysis to extract maximum intelligence from HEXAGON imagery. During his remarkable career, Lundahl enjoyed the confidence of Presidents Eisenhower, Kennedy, Johnson, and Nixon, as well as senior managers within the CIA and the DoD.

Arthur C. Lundahl

The National Imagery Interpretability Rating Scale (NIIRS)

In the early 1970s, as national collection and exploitation requirements expanded and became increasingly specific, it grew apparent that the Community needed a better measure for rating the quality of imagery—something that would provide a measurement of satisfaction of a stated imagery requirement. The measure that had been used since the first successful satellite mission was a rating of good, fair, or poor. Such a scale did not give the user or the collector very much information on the probability that a specific intelligence need had been answered. For example, was HEXAGON search imagery of good enough interpretability to detect a new Soviet antiballistic missile (ABM) facility?

The word "quality" has a different meaning for photoscientists than it has for collection system engineers. To balance this fact, a National Imagery Interpretability Rating Scale (NIIRS, pronounced "nears") was developed. The NIIRS substitutes the phrase "information potential for intelligence purposes" for "quality." The purpose of NIIRS was stated concisely as "to obtain from the photointerpreter a judgment as to the interpretability of an acquired image." As a result of the adoption of NIIRS in 1972, the Community users had a quick and accurate assessment as to whether or not a requirement had been met, and the collection manager (COMIREX) had a reliable manner in which to task the collector (NRO) or to remove tasking if the readout determined that required quality had been achieved.

One of the key needs for an improved imagery rating scale developed as a result of the first SALT and ABM treaties. A better method had to be found to report on the quality of HEXAGON search imagery and to determine if the imagery were good enough for detecting activities covered by the two treaties. Could it detect new or modified ICBM launching complexes, ABM launchings, and radars? Following each HEXAGON recovery, NPIC was tasked by COMIREX to NIIRS-rate all Soviet imagery; this information was incorporated into special SALT reports provided to US agencies and persons involved in treaty monitoring and negotiations. For example, a chart comprising the cumulative plotted NIIRS ratings of the Soviet Union was prepared annually for the President, the DCI, and elements of Congress. The chart showed in detail the areas covered and the quality of the coverage.

The NIIRS rating scale ranged from 0 (which meant that interpretability of the imagery precluded its use for photointerpretation) to 9 (which provided the highest interpretation capability). The following summary shows typical examples for the 10 NIIRS categories.

Rating Category 0

Interpretability of the imagery precludes its use for photointerpretation, due to obscuring, degradation, or very poor resolution.

Rating Category 1

Detect the presence of large aircraft at an airfield. Detect a launching complex at a known missile test range. Detect armored/artillery ground forces training areas.

Rating Category 2

Count accurately all large straight-wing aircraft and all large swept/delta-wing aircraft at an airfield.

Identify a completed Type III-C launching area within a known ICBM complex by road pattern/hardstand configuration.

Rating Category 3

Count accurately all straight-wing aircraft, all swept-wing aircraft, and all delta-wing aircraft at an airfield.

Detect vehicles/pieces of equipment at a SAM, SSM, or ABM fixed missile site.

Rating Category 4

Identify a fighter aircraft by type, when singly deployed.

Identify an SA-2 or CSA-1 missile by the presence and relative positions of wings and control fins.

Identify trucks at a ground forces installation as cargo, flatbed, or van.

Rating Category 5

Detect the presence of call letters/numbers and alphabetical country designator on the wings of large commercial/cargo aircraft (where alpha- numerics are 3 feet high or larger).

Identify an SA-1 transporter by overall configuration and details of chassis construction.

Identify a singly deployed tank at a ground forces installation as light or medium/heavy.

Rating Category 6

Identify a FAGOT or MIDGET aircraft by canopy configuration, when singly deployed.

Identify the following missile ground support equipment at a known strategic missile site: warhead/checkout van and fuel/oxidizer transporter.

Rating Category 7

Identify the pitot boom on a FLAGON aircraft.

Identify a strategic missile transporter/erector (fixed or mobile system) when not in a known missile activity area.

Rating Category 8

Identify on a FISHBED J aircraft the dielectric patch outboard on each wing leading edge and the horizontal tailplane tip spikes.

Identify the VHF antenna on the forward transit support assembly of an SA-4 transporter/launcher.

Rating Category 9

Identify on the appropriate model FISHBED aircraft: wing-flap actuator fairings, fairings in afterburner area above horizontal tailplane, pitot boom pitch-and-yaw vanes (when uncovered), and air dump port forward of canopy.

Identify a Mod-3 SA-2 missile by the canards (just aft of nose).

Weather Support—Key to HEXAGON Success

In spite of the fact that it carried more than 200,000 feet of film, HEXAGON can be characterized as a film-limited system because of the large number of requirements tasked to it and its long mission duration. Available film had to be utilized judiciously to take full advantage of long on-orbit life capabilities and to satisfy high-priority objectives. The key to effectiveness was good film management: one that produced maximum cloud-free imagery of each mission's stated requirements. Consequently, the accuracy of weather forecasts was critical to HEXAGON success. A large proportion of the priority search/surveillance areas tasked to HEXAGON were located on the Eurasian land mass. On any given day throughout the year, about 65 percent of this area was likely to be obscured by clouds. Without efforts to overcome the weather handicap, much of the coverage would have been obscured by clouds. The following graphics illlustrate this point. Graphic 4 shows the mean cloud- freeness for the month of January, and Graphic 5 shows the same data for the month of July. For both months there is less than a 40-percent chance of observing a point on the ground on any given day for the areas of primary intelligence interest.

Weather support comprised a continuous cycle during HEXAGON operations. Climatological data were used extensively during the mission planning stage to help in selecting the launching date and time and as an input to mission planning software that affected such factors as requirement weights (priorities), film allocation, weather thresholds, and requirements satisfaction goals. Climatology also played an important role in on-orbit operations. For example, if the probability of successful coverage of South China was highest in December and January, marginal opportunities for photography could be passed up in August or September to concentrate collection efforts in the months with a higher probability of success.

On-orbit weather support was provided by Global Weather Central (GWC) from its facilities at the Strategic Air Command (SAC) headquarters in Omaha, Nebraska. GWC was a component of the Air Force's Air Weather Service. Accurate on-orbit forecasts and verifications were primarily dependent on weather satellites developed by the Air Force in the "white" program 417, funded by the NRO (see Graphic 6). Optimum support was provided by a morning scout satellite, used for forecasts, and an afternoon satellite, used to provide weather verification of areas imaged earlier. In actual practice, due to launching problems or unexpected on-orbit failures, it was not always possible to have both morning and afternoon weather satellites in action.

A second key element in the forecasting process was the information reported by thousands of weather stations scattered around the world, including in the Soviet Union. Broadcasts from Soviet stations were intercepted by SIGINT collection means and relayed to GWC. Under ideal conditions, weather forecasts for an upcoming HEXAGON pass could be based on weather data about two hours old. Weather verification data for areas imaged could also be as fresh as about two hours.

How good was GWC forecasting? As noted above, two-thirds of the Eurasian landmass is cloud-covered at noon on any day of the year, but HEXAGON's cloud-free return for the area consistently ranged in the 70- to 85-percent range. This despite the fact that the Intelligence Community often levied requirements which, because of their high priority, had to be attempted under weather conditions that were forecast to be poor.

To improve the mission effectiveness further, a quick check of the GWC weather verification was made after each "bucket" recovery (except the fourth) by rushing a copy of the film to Washington where the Defense Mapping Agency, which had the in-house film-delineation resources, produced a cloud-cover readout that was then converted into world aeronautical grid (WAG) cells, the requirements accounting measure used by GWC. The readout was quickly passed back to the operator to update the mission requirements file.

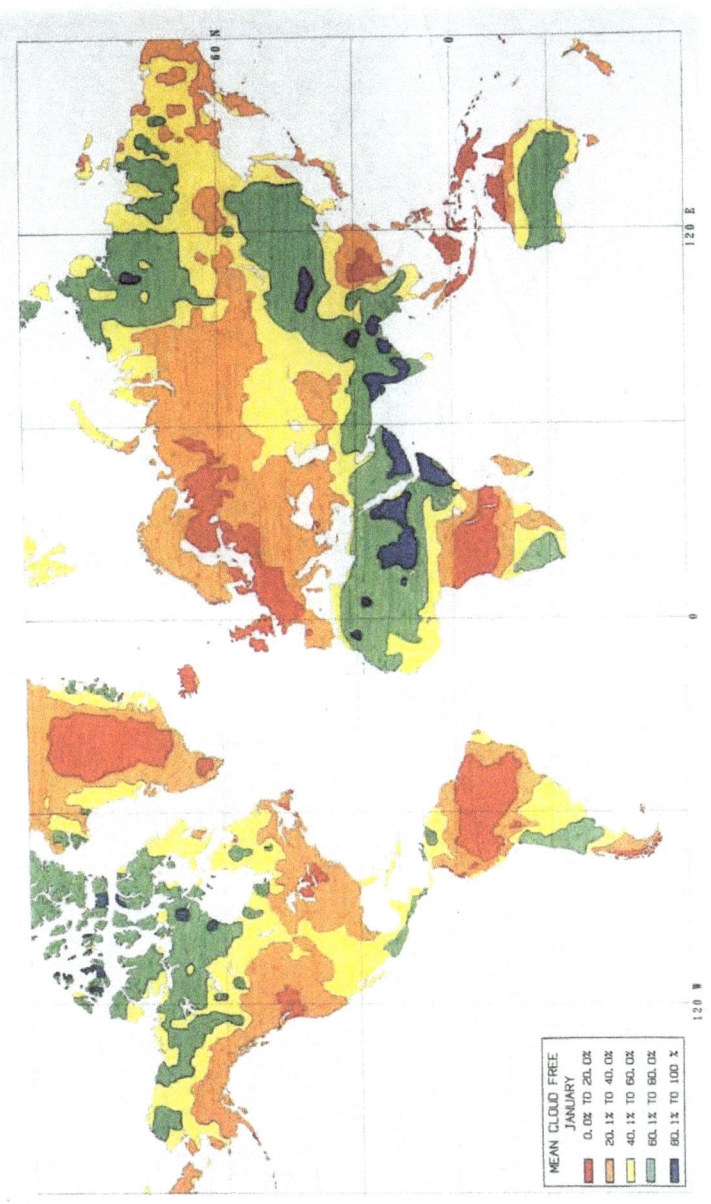

Graphic 4. Mean Cloud-Free Areas of the World in January

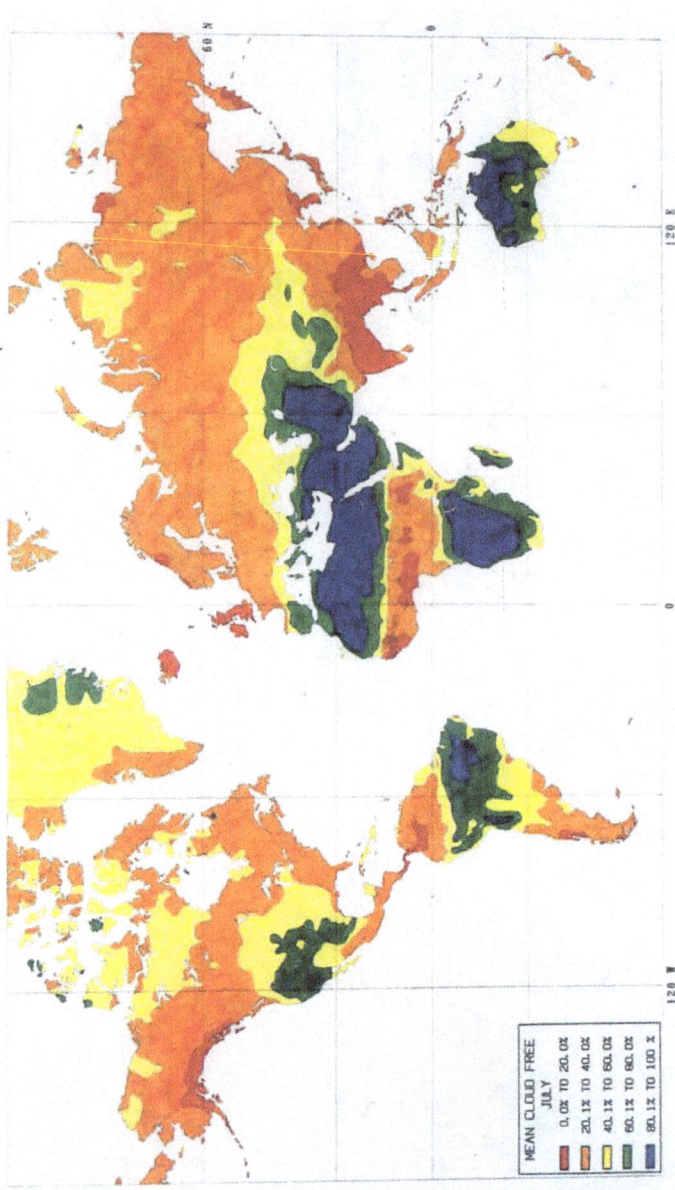

Graphic 5. Mean Cloud-Free Areas of the World in July

WEATHER SATELLITES - PROGRAM 417
OBTAIN COVERAGE OF DENIED LAND AREAS

SYSTEM ELEMENTS

- BOOSTER — THOR/BURNER II
- RCA SPACECRAFT
- ORBIT CONTROL — SPIN STABILIZED
- RCA VIDEO SYSTEM
- RCA/BARNES IR SENSOR

PAYLOAD DATA

- VIDEO RESOLUTION — 1.0 NM
- CLOUD TOPS MEASURED TO ±1000'
- DATA RECORD AND READOUT

ORBITAL PARAMETERS

- INCLINATION — ~98 DEG
- ORBIT — ~450 NM CIRCULAR
- LIFETIME — ~9 MO

Graphic 6. Program 417 Weather Satellites

~~SECRET~~
~~NOFORN-ORCON~~

HEXAGON Collection Requirements Formulation

As the HEXAGON hardware development improved, so did the Community's requirements review and definition. For example, an in-house CIA assessment in the mid-1960s defined first-priority objectives for the 1970s as follows:

Priority objectives of the photo-satellite effort in the 1970s will be the first to monitor the strategic capabilities of the USSR and Communist China. Satellite photography is essential for monitoring such major aspects of these capabilities as:

A. The deployment and mode of operation of strategic forces, both offensive and defensive.

B. The research, development, and testing of weapons systems related to strategic systems.

C. The production, testing, and stockpiling of fissionable materials and nuclear weapons.

D. The composition, strength, disposition, order of battle, readiness, and combat roles of general-purpose forces.

E. The capacity and operating status of the industrial and logistics establishment supporting military forces.

Throughout its lifetime, HEXAGON was to provide a significant input toward meeting these priority objectives.

At the time of its introduction, HEXAGON had the unique capability to satisfy three major intelligence needs: search, surveillance, and MC&G.

Evolution of HEXAGON Broad-Area Search Requirements

Broad area search (BAS) imagery collection and exploitation are conducted worldwide for the purpose of timely detection of previously unknown installations or activities associated with any current intelligence problem. The primary objective is to eliminate surprise and increase confidence in the overall intelligence production process. HEXAGON satellite imagery intelligence was uniquely capable of accomplishing this objective by providing a permanent accountable record of direct evidence which confirmed or denied the presence of new activity in large, contiguous geographic areas.

Search requirements for HEXAGON evolved from those defined for CORONA and were revised frequently to meet newly recognized intelligence needs. The initial HEXAGON-defined search mission was included in the 1966 USIB[274] system definition and was stated as follows:

> **Search Mission.** KH-9 should have the capability to provide stereoscopic, cloud-free (about 90 percent) photography of about 80 to 90 percent of the built-up areas of the Sino-Soviet bloc (approximately 6.8 million square nm) semiannually and should provide similar coverage of about 75 percent of the undeveloped areas (2.8 million square nm) annually (see Graphic 7). It should be noted that this requirement differs from that approved by USIB on 19 March 1965 (USIB-D- 41.14/229; COMOR-D-13/43) and that it is based on the results obtained and general satisfaction with search coverage acquired over the last 18 months with the KH- 4. In addition to search of the Sino-Soviet bloc, KH-9 should provide the capability to acquire coverage of contingency areas in other parts of the world on demand.
>
> Present areas requiring this coverage are Indonesia, the Middle East, Southeast Asia, and parts of North Africa. We do not expect this requirement to exceed 3 million square miles per year.

For the first time, the Community recognized the collection efficiency in proposing a high probability of "detection," that is 90 percent cloud-free photography, 80 to 90 percent of the built-up areas, and 75 percent of the undeveloped areas. The application of probability rules in defining collection requirements was to play an important role in defining future search requirements. To ensure that HEXAGON search/surveillance requirements were compatible with the mission-planning and targeting software under development, the Community updated its collection requirements[275] in 1969. The new requirements reaffirmed the basic structure outlined in 1966, amplified major elements within it, and introduced several new features.

A major innovation was adoption of the 1:50,000 WAG cell, an area averaging about 12 by 18 nm, as the unit of accounting for defined area search requirements. The WAG system, used universally for aeronautical navigation, already had been adopted by the NRO as a tool for use in managing collection operations. This system permitted the Community to delineate and classify the various categories of search and surveillance areas to a much higher degree than had been possible. The WAG cell was used in the HEXAGON program for four purposes: delineation of requirements, tasking requirements to the NRO, developing NRO targeting software, and reporting the exploitation/requirements satisfaction process. For example, standing and special search requirements satisfaction reporting was accomplished on the basis of the percentage of WAG cells imaged satisfactorily during the specified collection period. (An illustration of the WAG cell system is shown in Graphic 8.)

~~SECRET~~
~~NOFORN-ORCON~~

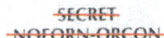

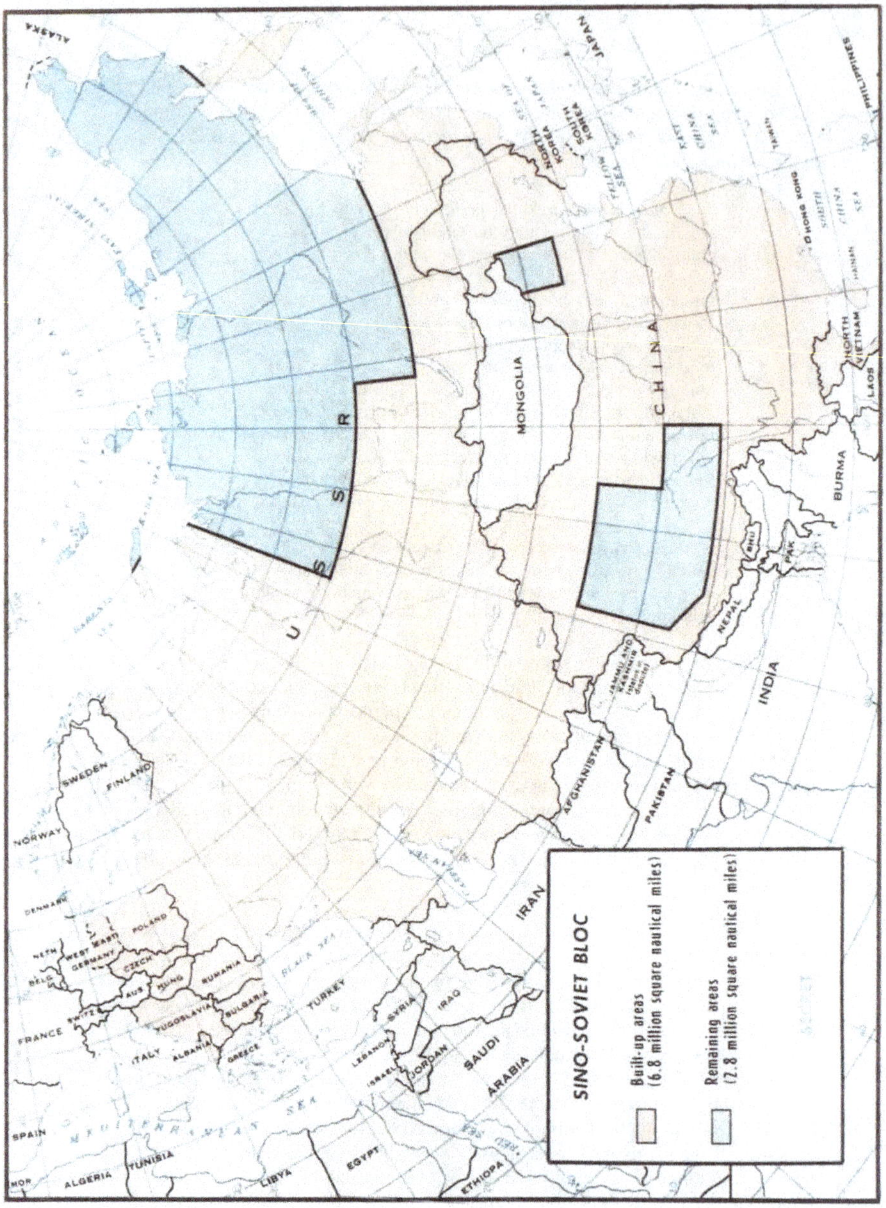

Graphic 7. Map Which Shows First Breakout Of Built-Up And Undeveloped Regions

~~SECRET~~
Handle via
BYEMAN-TALENT-KEYHOLE
Control Systems Jointly
BYE 140002-91

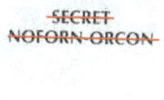

Graphic 8. World Aeronautical Grid (WAG) Chart

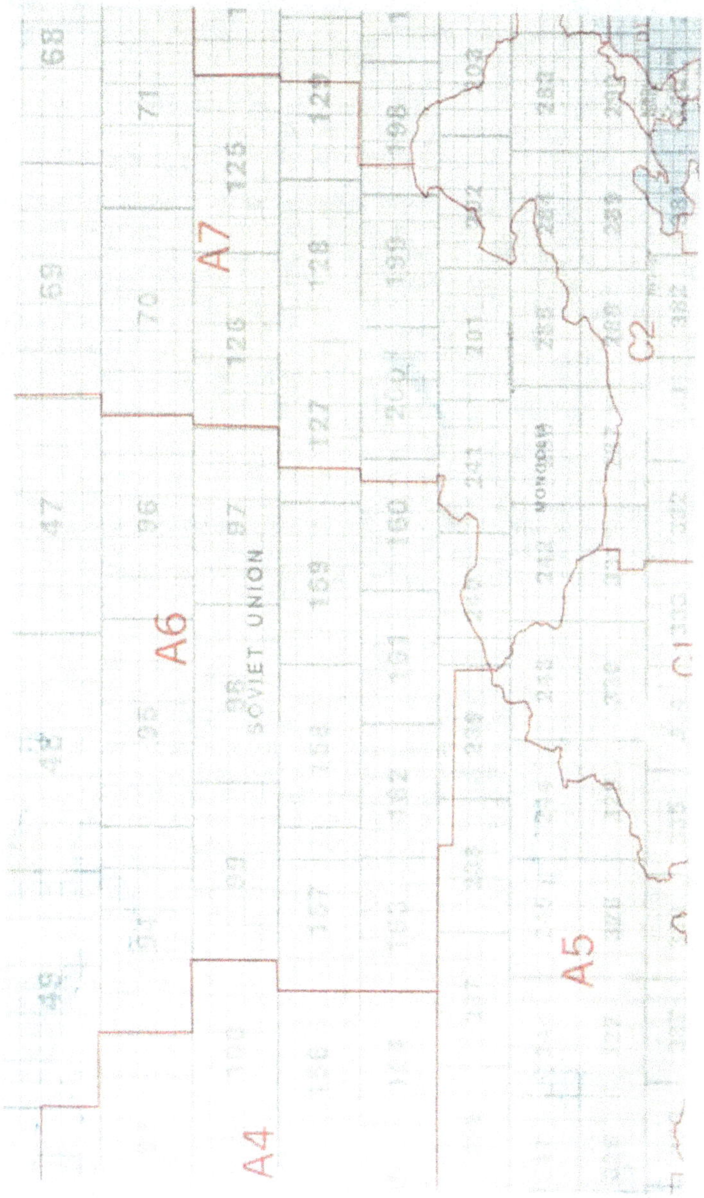

Coverage of the year-round, all-weather transportation routes leading to most sites of significance to the Intelligence Community was recognized as a key requirement, and as a consequence, the built-up areas were defined in terms of proximity to transportation. In the absence of any techniques for precisely defining the term of "proximity to transportation," the Community retained the figure adopted in 1961—15 nm. Under this rule, if any portion of a WAG cell fell within 15 nm of a transportation artery, the entire cell would be counted as part of the built-up area.

Another feature was the precise delineation of 107 target clusters and the specification that they be covered quarterly with 25 percent of the clusters photographed 85 percent cloud-free or better and the remainder 70 percent cloud-free. The objective of this coverage was search as well as surveillance; the clusters were recognized as the most likely areas for new targets to appear since "new installations of military importance are frequently located near or within facilities of similar nature. . . ." These more precise delineations of the cluster and built-up regions led to a reduction of their combined size from 6.8 million to about 5.1 million square nm.

The actual delineation of the target clusters and built-up regions by WAG cell on the basis of major lines of communication (LOCs) was a very large and complex task performed by CIA's Office of Basic and Geographic Intelligence (OBGI). Using large-scale maps overlaid with WAG grids, each WAG cell that fell within 15 nm of a major LOC was identified and stored in a computer database for later use in preparing graphics that delineated the new requirements (see Graphic 9).

Standing Search Delineations in the 1970s

The initial delineation of WAG cells for the target clusters, built-up regions, and undeveloped areas as called for in the 1969 amplification of the KH-9 requirement was completed in early 1970. Follow-on detailed studies were conducted on the location of significant all-weather transportation routes and changing criteria for the designation of target clusters. These studies led to revisions in the delineations, which were promulgated in mid-1971 and in the fall of 1972. In early 1973, nearly 1.5 million square nm of the most inhospitable of the undeveloped areas were split off and designated as remote regions. At least 80 percent of each of the three remote categories identified for the USSR, China, and Mongolia was to be imaged every 18 months.

In mid-1976, a fifth coverage category, topographically unsuitable, was created through the subdivision of the regions previously designated as remote.

In 1977, COMIREX provided additional collection guidance which embraced a still more sophisticated breakdown of collection frequencies and a further elaboration of area delineations. A major impact on the new delineations was caused by stringent application of the "target proximity theory." Studies in 1975[276] had shown that important targets tend to be in clusters. ▓▓ percent of active COMIREX targets in Eurasian Communist countries and the Middle East were within ▓▓ of another target. Even in the 75 percent of the land area where significant cultural activities were sparse or absent, more than half of the relatively few intelligence targets present were within ▓▓ of one another.

The reason targets tend to cluster is logical and well understood. Many types of military units, for instance, must be located in border regions to facilitate defensive and offensive operations. Such units and most important functions of military and intelligence significance require logistic support and thus are positioned on or near major LOCs. In many instances, new military facilities are located near existing ones to avoid the expense of having to develop from scratch the logistics, housing, communications, and other support bases required to maintain permanent operations. Heavy industry is concentrated in regions having not just the necessary mineral resources but also adequate supplies of labor, water, and electric power. Defense plants tend to be clustered near key suppliers and/or pools of skilled labor. Thus, the tendency of intelligence targets to be clustered was recognized and taken advantage of in the development of imaging requirements and collection strategies.

The seven newly delineated categories were defined as follows:

- **Clusters.** The most target-dense portions of the delineated regions. Although constituting only about ▓▓ percent of the total land area covered by delineated standing requirements, these clusters contained nearly ▓▓ percent of the targets then active in the COMIREX target deck.

- **LOC Target Areas.** Part or all of each WAG cell included in this category was both within ▓▓ of at least one active COMIREX target and within ▓▓ of an all-weather LOC. Altogether, this category covered about ▓▓ percent of the land within the delineated requirements area and contained about ▓▓ percent of the active targets.

- **Remaining Target Areas.** The clusters and LOC target areas together contained about ▓▓ percent of the active COMIREX targets, yet comprised only about ▓▓ percent of the total land area. The remaining targets were widely scattered. This category was composed of WAG cells either largely or completely within ▓▓ of at least one of those scattered targets.

- **Other LOC Areas.** Part or all of the area within each cell in this category was within ▓▓ of an all-weather LOC, but none of the cells was within ▓▓ of an active target. It was estimated at the time that about ▓▓ percent of future targets would be located within the ▓▓ percent of the land area contained in this category.

- **Undeveloped Areas.** Cells in this category, which comprised about ▓ percent of the land area, were at least ▓ from all-weather LOC and at least ▓ from any known target and were served by only rudimentary transportation systems.

- **Remote Areas.** In this category were cells more than ▓ from any COMIREX target in regions without meaningful transportation arteries. About ▓ percent of the land area fell within this classification.

- **Topographically Unsuitable Areas.** Those regions marked by towering mountain ranges, swamps, lakes, and glaciers and deemed as highly unlikely areas to support new activities or targets of national interest to the Intelligence Community.

The 1977 guidance also addressed quality, mode, and frequency of coverage considerations. In the case of target clusters, no imagery poorer than NIIRS 4 was to be counted toward requirement satisfaction. For the remaining six delineated categories, no imagery poorer than NIIRS 3 quality was to be counted, and at least ▓ percent of the imagery counted had to be rated NIIRS 4 or better.

For the first four categories above, stereoscopic coverage was required. For the remaining categories, stereoscopic coverage was preferred but monoscopic was acceptable. The coverage periods for each of the seven delineated categories were as follows:

Category	Coverage Frequency in Months
Clusters	2
LOC Target Areas	4
Remaining Target Areas	6
Other LOC Areas	9
Undeveloped Areas	12
Remote Areas	18
Topographically Unsuitable Areas	24

The 1977 requirements statement was the last formal requirements revision to affect HEXAGON operations. A 1979 "BAS Statement of Requirements for Mid-1980s Planning" confirmed that the existing (1977) requirements would continue to constitute the primary intelligence search needs. It also formally recognized Third World areas that had regularly been tasked to HEXAGON missions as ad hoc (or special) requirements. These were divided into secondary and tertiary search areas with defined collection frequencies and quality. Finally, in 1983, a new requirements statement "BAS Requirements for the Mid-1980s and Beyond" was undertaken to define BAS requirements for the follow-on search system. (Graphic 10 shows an example of BAS requirements.)

Age of Search Imagery

Graphic 11 shows the desired age distribution of search imagery that would provide a high confidence that any new activity of intelligence significance would be detected within a reasonable time period. For example, ▇ percent of the WAG cells should be collected by the time that half of the stated collection frequency has expired, and ▇ percent should be collected by the end of the requirements period. For example, to satisfy the undeveloped area search ▇▇▇▇ requirement, ▇ percent of the area should be seen within any ▇▇▇▇ period and ▇ percent seen within any ▇▇▇▇ period. Generally, HEXAGON met the need for those requirements that had longer periodicity; it sometimes fell short of meeting short-period requirements, depending on the frequency of missions flown.

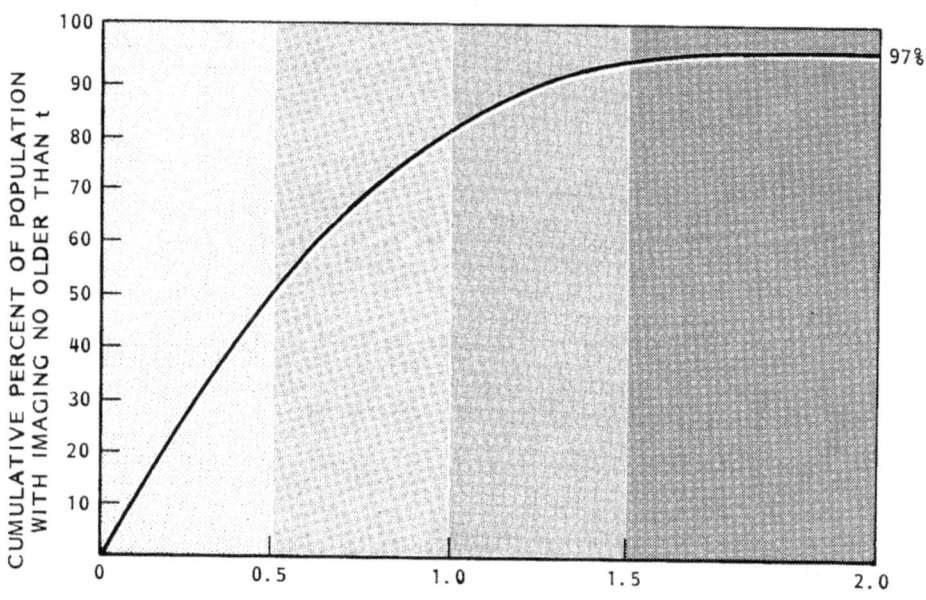

Graphic 11. Cumulative Age Distribution

HEXAGON Search Capabilities

Search is conducted worldwide for the purpose of timely detection of previously unknown installations or activities associated with any current intelligence problem.

HEXAGON's search mission was defined, in part, as follows:

> **Search Mission.** KH-9 should have the capability to provide stereoscopic, cloud-free (about 90 percent) photography of about 80 to 90 percent of the built-up areas of the Sino-Soviet bloc (approximately 6.8 million square nm) semiannually and should provide similar coverage of about 75 percent of the undeveloped areas (2.8 million square nm) annually. In addition to search of the Sino-Soviet bloc, KH-9 should provide the capability to acquire coverage of contingency areas in other parts of the world on demand.
>
> Present areas requiring this coverage are Indonesia, the Middle East, Southeast Asia, and parts of North Africa. We do not expect this requirement to exceed 3 million square miles per year.

HEXAGON's ability to satisfy the stated search requirement varied throughout the program's life and was dependent primarily on the actual launching rate and the numbers of special requirements tasked to each mission.

Graphic 12 shows the level of satisfaction of the standing search requirements maintained over the three-year timespan from 1974 to 1976. This graphic does not, of course, reflect HEXAGON's responsiveness to special search requirements that were tasked to each mission. For example, considerable resources might be expended in searching all probable ABM deployment areas. This could have a very high priority from an intelligence standpoint, but would contribute little to satisfying standing search requirements because most ABM special search areas were in easier-to-satisfy, longer-period search areas. Another example of HEXAGON's responsiveness to special search needs is shown in Graphic 13, which depicts a successful effort to search essentially all of ▓▓▓▓▓ on mission number 1213) in order to confirm/negate ▓▓▓▓▓

Three different summary tables of Mission 1217 requirements satisfaction are provided to illustrate the tremendous area coverage capacity of a single HEXAGON mission. Table 1 depicts the level of effort against standing and special search requirements tasked to this mission: over ▓▓▓▓▓ square nm attempted and over ▓▓▓▓▓ square nm of unique area attempted, over ▓▓▓▓▓ of which was cloud-free. Table 2 shows the effort against selected special search requirements ranging in size from over 800,000 square nm to about 8,000 square nm. Each of these special requirements shows a high level of satisfaction, except for the Colombia requirement. Colombia is located in one of the poorest weather areas in the world. Table 3 is a compilation of the total mission coverage by mission increments and geographic areas.

Requirement	Area (million sq. nm)	Frequency*
●—— Built-up	3.9	
—— Undeveloped	3.5	
—·— Remote	2.7	
······ Topographically Unuseable	.5	

*Requirement is to achieve an age distribution of coverage so that ▇ of the respective areas have been seen at the stated frequency

Note: Status is shown as of the end of each month.

Note: The topographically unuseable category resulted in a further refinement of the collection requirements

Graphic 12. Status of Search Coverage

**Table 1
Mission 1217
Intelligence Search Summary (Standing and Special)
By Geographic Area* (TS/TK)**

*Accomplishments calculated on the basis of 3- by 3-nm WAG subcells, mono and stereo coverage. (S/TK)
†This does not address repetitive requirements, for example:

~~SECRET~~
~~NOFORN-ORCON~~

Table 2
Mission 1217
Selected Intelligence Special Search*

*Calculated on the basis of 3- by 3-nm subcells. (S/TK)
†Stereo coverage only. (S/TK)

Table 3
Mission 1217
Intelligence and MC&G Worldwide Coverage*
By Recovery Vehicle

Mission RV			
1217-4			
1217-3			
1218-2			
1217-1			
Total			

*Calculated on the basis of 3- by 3-nm WAG subcells imaged cloud-free. (S/TK)

†Represents the unique area imaged *within* each RV; redundant coverage as between RV is therefore *included* in the figure. (S/TK)

‡Represents the unique area imaged for the entire mission; redundant coverage whether between RVs or within an RV is excluded from the figure. Unique coverage is computed from the most recent coverage, i.e. 1217-4, 1217-3, etc. (S/TK)

By Geographic Area* (TS/TK)

Geographic Delineation		
Total		

*Square nautical miles (sq nm) calculated on the basis of 3- by 3-nm WAG subcells. Figures for gross coverage include both mono and stereo coverage. (S/TK)

†This includes the total area associated with the geographic delineation regardless of whether the area was actually tasked to Mission 1217 for collection. (S/TK)

HEXAGON Surveillance Capabilities

Surveillance is the periodic coverage of installation, objects, or activity for the purpose of updating information previously obtained.

In the 1966 USIB requirements definition for HEXAGON,[274] the surveillance mission was defined as follows:

> **Surveillance Mission.** In recognition of the capability of KH-9 to obtain high-resolution area coverage when meeting the specifications above, we believe it appropriate to specify frequency of coverage in terms of surveillance of geographic areas representing target clusters rather than in terms of surveillance of individual point targets. Based on target distribution, we have identified about ▓ clusters ranging in size up to ▓▓▓▓ mile areas in which approximately ▓ percent of current targets are located. As new targets are added to the list, it is expected that the great majority will also fall in these same clusters. Although the bulk of these areas are located within the Sino-Soviet bloc, several of similar size fall outside this area. These target clusters, each of which contains a variety of target category types, should be considered dynamic and therefore subject to change as experience with KH-9 is acquired. For planning purposes, however, we believe that surveillance of about ▓ percent of these areas quarterly should be accomplished, especially since the KH-8 high-resolution spotting system can be employed to round out coverage or to obtain additional coverage as may be deemed necessary.

HEXAGON was capable of meeting a significant proportion of the Community's surveillance requirement. The quality of HEXAGON imagery was adequate to satisfy a large proportion of stated intelligence needs. A 1968 COMIREX assessment,[277] for example, noted that HEXAGON's planned resolution would satisfy ▓ percent of projected surveillance requirements from an image-quality standpoint. Table 4 illustrates the tremendous number of national interest targets a typical HEXAGON mission was able to image (in this case, ▓ percent of all active COMIREX targets). In addition to these targets, thousands of additional targets carried in the DoD Bombing Encyclopedia were also imaged and reviewed by DoD organizations.

Actual HEXAGON experience confirmed the 1968 COMIREX projection. Table 5 shows the relationship of COMIREX standing surveillance requirements to quality (NIIRS), and Graphic 14 shows the typical NIIRS distribution of COMIREX targets on a mission. The actual mission results clearly demonstrate HEXAGON's capability to meet a high proportion of the standing surveillance requirements in terms of quality.

Table 4
Mission 1217
COMIREX Point Target Coverage
Geographic Summary (TS/TK)

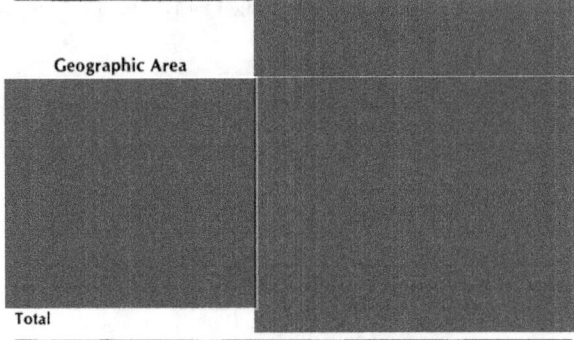

*The figures given here reflect all COMIREX targets active for collection as of 30 November 1982. (S/TK)

Category Summary (TS/TK)

*The figures given here reflect all COMIREX targets active for collection as of 30 November 1982. (S/TK)

Table 5
Relationship of Standing Imagery Surveillance Requirements

Standing US Needs for Imagery-Related Intelligence		Summary of Surveillance Requirements				
Major Subject Area	Aggregated Intelligence Topics	Total Target* Requirements	Mode	Quality NIIRS†	Sampling Frequency	Daily Demand‡
Political			M/S	3-6		
			M	5		
			M	4-6		
Economic			M/S	4-5		
			M/S	4-6		
			M/S	4-5		
Military			M	3-6		
			M/S	5-6		
			M/S	3-5		
			M/S	4-6		
			M/S	3-6		
			M/S	5-6		
			M/S	4-6		
			M/S	4-6		
			M	4-6		
			M	4-5		
			M/S	4-5		
			M	4-6		
			M/S	4-6		
			M/S	5-6		
			M/S	4-6		
			M/S	5-6		
Special Subjects			S	5		
			M	4-6		
			M/S	4-6		
	Total					

*Any specific target may be grouped in one or more of the broad intelligence problems; hence, total requirements represent this multiplicity of intelligence contributions.

†The National Imagery Interpretability Rating Scale (NIIRS) is a graduated scale designed to judge quality for intelligence purposes. There are 10 levels on a geometric progression ranging from useless (NIIR 0) to the best imagery systems (NIIRS 9).

‡Represents the demand on a daily basis of clear target images at specified quality and mode.

Graphic 14. NIIRS Distribution and Cumulative Distribution of Unique Comirex Targets Covered on KH-9 Mission 1210

HEXAGON and Third World Countries

HEXAGON contributed to US knowledge of Third World countries more than any other system before or since. It was through HEXAGON that we became aware of transportation, industrial facilities, and military deployments in Third World countries. It was from HEXAGON that we located ▮▮▮

Color Imagery

There were predictions that color would add much to the intelligence information collected by HEXAGON; however, with the exception of its contributions to economic intelligence, this did not prove to be the case. A variety of color and near infrared films were flown on HEXAGON. A number of special analyses of color imagery were accomplished but these special films never produced any significant intelligence items that could not have been observed on black-and-white film. In fact, on several occasions, the color films' poorer resolution inhibited photointerpreter readout.

There were some applications of HEXAGON product to economic intelligence for which color was useful. This was particularly true in coverage of grain production and in oil- and mineral-area potential evaluation.

HEXAGON Mapping, Charting, and Geodesy (MC&G) Capabilities

In 1966 the USIB approved the following statement of HEXAGON's MC&G requirements:

> **Mapping and Charting.** For KH-9 photography to be used directly in the preparation of maps and charts, it must contain the strong geometry required to meet the horizontal and vertical accuracy for large- and medium- scale maps and charts of which the most demanding is the large scale (1:50,000) topographic maps. These maps require a relative horizontal accuracy of 85 feet and a vertical accuracy of 16 to 33 feet over a distance of 10 to 20 miles. An accurate photogrammetric control network extending 500 miles in any direction within specified regions is essential for the development of an orderly production of coordinated series of maps and charts. KH-9, in addition to providing search/surveillance as stated above should also provide coverage of about 7 to 10 million square miles of the free world each year. This requirement usually can be satisfied by one-time coverage supplemented by re-coverage of relatively small areas (see COMOR-D-13/65 for additional statement of requirements).

It would be difficult to dispute an argument that HEXAGON was the ultimate design for a mapping system. Certainly, in the forseeable future, there is no planned replacement system that provides the simultaneous coverage of large, contiguous areas of the earth at large scale and at required geodetic accuracies.

The Defense Mapping Agency (DMA) and its predecessor organizations (CIA and other government agencies that produced maps and charts) were almost solely dependent on HEXAGON for mapping source materials. Source materials for MC&G products were provided initially by the main camera, mission number 1201 through 1204; a combination of the main camera and the 12-inch focal length mapping-camera system (MCS) on mission numbers 1205 through 1216; and the metric main camera system on mission numbers 1217 through 1219.

In general, mapping products are generated at various scales for air, ground, sea, and space operations, and for intelligence and military planning. The geodetic data derived from satellite imagery provides the military with tens of thousands of accurate point locations needed for operation of strategic and tactical weapon systems. MC&G requirements can be divided into three categories: point-target requirements used to update information files; broad area coverage (non-metric) for original map compilation and revision; and broad area coverage metric requirements for original compilation requiring accurate point positioning. Each HEXAGON mission contributed significantly to these requirements.

HEXAGON satisfaction of DMA's stated MC&G requirements was a function of the number of HEXAGON missions flown annually and the proportion of each mission's film that could be allocated to mapping needs. Even the reduced launching schedule (toward the end of the program) satisfied extensive mapping requirements. In one sense, any clear imagery collected for any purpose has potential MC&G utility since a new requirement in any area of the world can develop at any time. Table 6 illustrates the mature HEXAGON Program contribution to MC&G requirement satisfaction on a single mission (number 1217). It shows that over 12 million square nm of clear imagery either satisfied a mapping requirement or was adequate to satisfy future or potential requirements. This 12 million square nm represents about 60 percent of this mission's total cloud-free imagery.

THE HEXAGON STORY

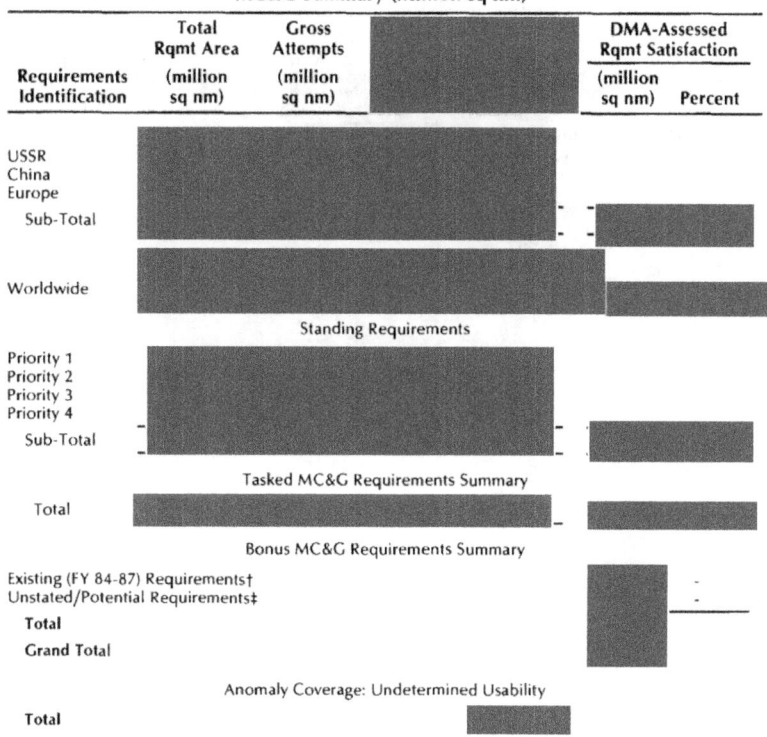

Table 6
Mission 1217
MC&G Summary (Million sq nm)*

*Calculated on the basis of 3- by 3-nm subcells. Some of these subcells will be retasked on subsequent missions to provide 90-percent cloud-free contiguous coverage of their related 12- by 18-nm WAG cell. (S/TK)

†This coverage assessed by DMA to be usable to support existing MC&G requirements that were not identified for tasking this mission. (S/TK)

‡This coverage assessed by DMA to support future but yet unstated MC&G production requirements. (S/TK)

NOTE: The total DMA requirements represent all current outstanding requirements—not total mapping requirements expected to be satisfied on a single mission.

Epilogue

In this final volume in the series of documents that provide historical perspective on the film-return programs developed by the National Reconnaissance Program, it is appropriate to review comparative imagery examples from each of these programs. The following images of the US Capitol (Graphics 15, 16, 17, and 18) at the same magnification graphically illustrate the improvement in quality through the evolution of the film-return systems. Also shown are a graphic of Launching Site 1 at the ███████████████████████████████████████ showing comparative CORONA, GAMBIT, and HEXAGON imagery (Graphic 19) and a comparative graphic of the "███████████" that demonstrates each system's capability to accept magnification of ███████ or greater (Graphic 20).

Imagery Illustrations

The following examples of HEXAGON imagery are included to convey a sense of some of the intelligence problems that HEXAGON helped to solve. In addition, some examples of the best imagery from the mature system are included to provide an appreciation of the very high-quality imagery achieved by the HEXAGON system.

- **Arms Limitations Agreements.** US agreement to the first Strategic Arms Limitation Talks (SALT) Treaty in 1972 was made possible by the ability of the satellite reconnaissance program to monitor Soviet research and development, production, and deployment of strategic offensive and defensive weapons systems. During its lifetime, HEXAGON played the key role in monitoring such activities and deployments. Graphic 21 depicts the signing of the first SALT agreement between President Nixon and Soviet Chairman Leonid Brezhnev in 1972. The US Government delayed the signing of this agreement until the quality of HEXAGON imagery could be confirmed through analysis of the first mission's imagery. Graphic 22 shows a typical HEXAGON 90-degree frame of imagery encompassing an area of about ▓▓▓ square nautical miles (sq nm). Considering that the USSR encompasses an area of almost 7 million sq nm and the mature HEXAGON system would image about ▓▓▓▓ of this area cloud-free on a typical mission, the task of the National Photographic Interpretation Center (NPIC) to assess this imagery for SALT verification purposes was significant. For example, Mission 1217 covered over ▓▓▓▓ sq nm of the USSR uniquely. This would equate to over ▓▓▓ frames of imagery of the size shown in Graphic 22—a tremendous search task for NPIC to accomplish in a timely manner. Graphic 23 shows destroyed SS-7 silos at the Perm ICBM complex in the Soviet Union. These older ICBM launching facilities were destroyed to stay within the allowed numbers of launch facilities as newer ICBMs were brought into the inventory. Graphic 24 shows the ability of HEXAGON to provide total coverage of a specific SALT-related issue, SS-7/SS-8 dismantling. It shows complete HEXAGON coverage of all launchers on two successive missions as compared with partial coverage on two successive GAMBIT missions.

Graphic 21. Signing of SALT I Agreement 1972

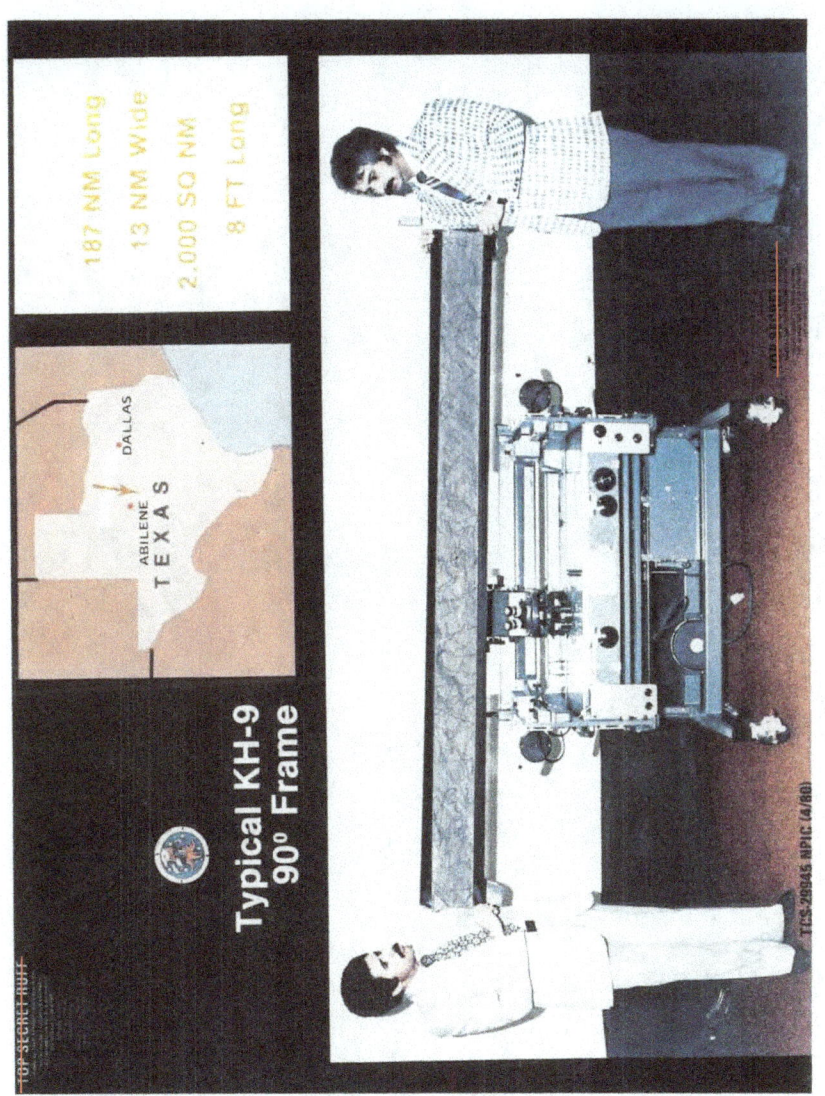

Graphic 22. Typical KH-9 Frame

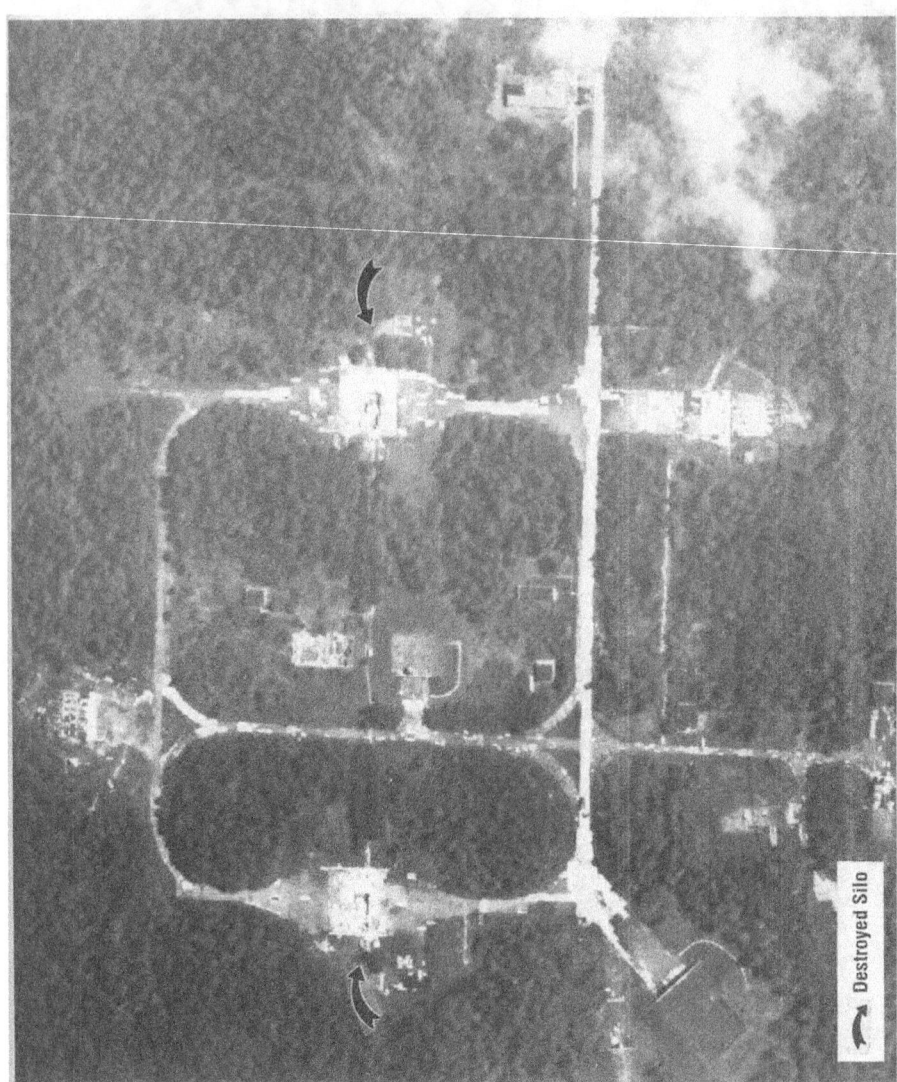

Graphic 23. Perm ICBM Complex – Destroyed SS-7 Silos – Mission 1215 40X

Mission	Launchers	Dates
KH-8	150	22 March - 17 May 76
KH-9	209	5 December 75 - 29 March 76
KH-8	146	16 September - 5 November 76
KH-9	209	9 July - 9 December 76

Graphic 24. SS-7/SS-8 ICBM Launch Site Dismantling

- **Detection**. Detection of new activities or facilities of intelligence interest was one of the primary tasks of HEXAGON. Graphic 25 shows re-coverage of the Mishelevka phased-array radar, which had been detected on an earlier mission. Graphic 26 shows the ▬▬▬▬▬ which was detected on an earlier HEXAGON mission. HEXAGON's broad-area coverage capability, coupled with the COMIREX-defined requirement for periodic coverage of areas of intelligence interest, provided high confidence that new installations and activities of intelligence significance would be detected early in the construction phase.

Graphic 25. Mishelevka Phased-Array Radar, USSR – Mission 1218-3 20X

- **Military Forces Order of Battle Information.** The ability of HEXAGON to furnish high-quality imagery of military installations during each mission increment allowed US intelligence analysts to develop and maintain very accurate order-of-battle information on Soviet, Warsaw Pact, Chinese, and other nations' forces. Entire Soviet military districts could sometimes be imaged on a single mission, providing current and accurate force-structure assessments. The following graphics illustrate this capability.

    Graphics 27 and 28 depict Soviet army barracks illustrative of regular and frequent coverage of Soviet army facilities by HEXAGON to maintain current ground order-of-battle information.

    Graphics 29 and 30 illustrate naval order-of-battle information available from HEXAGON imagery.

    Graphic 31 shows coverage of a Soviet BACKFIRE base, a high-interest strategic target that required regular coverage by HEXAGON.

Graphic 29. Vladivostok Naval Base, USSR — Mission 1213-3, BOX

Graphic 30. Leningrad Shipyard, USSR – Mission 1218-1 40X

Graphic 31. BACKFIREs at Belaya Airfield, USSR – Mission 1218-3 40X

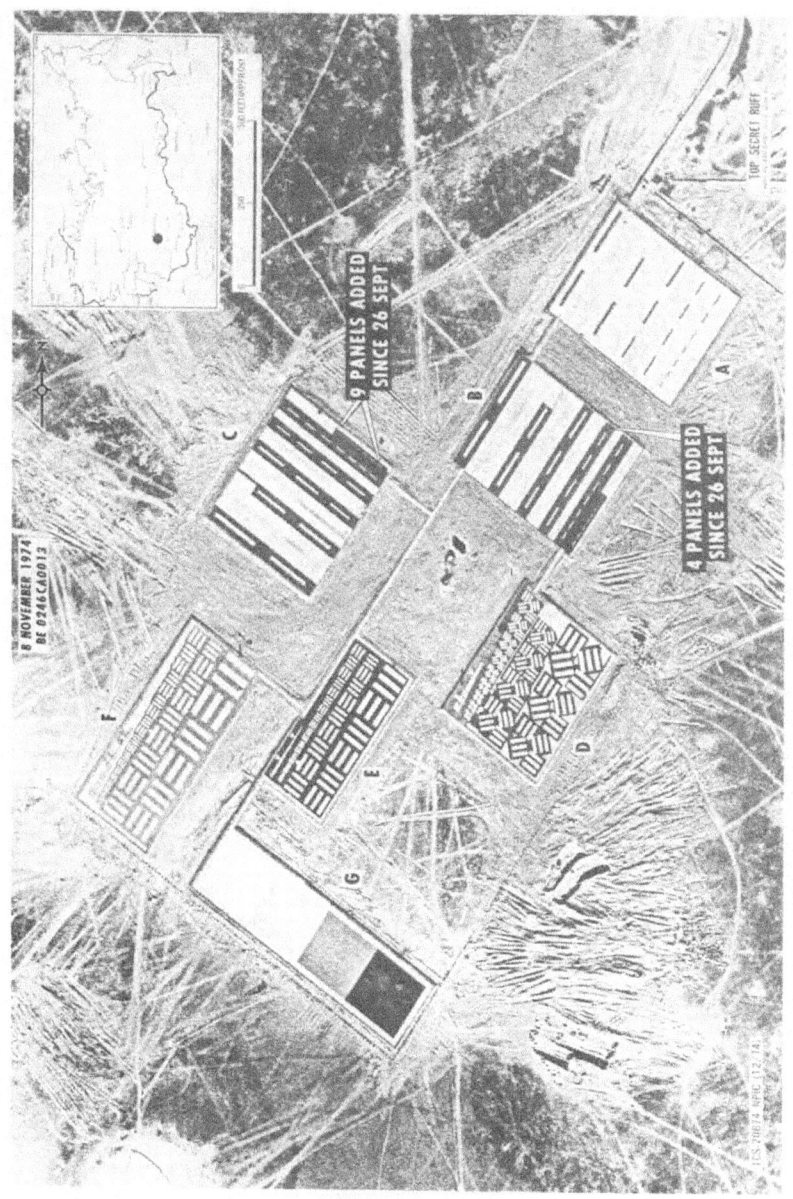

Graphic 44. Reconnaissance Target Area, Tyuratam Missile/Space Test Center SSM, USSR

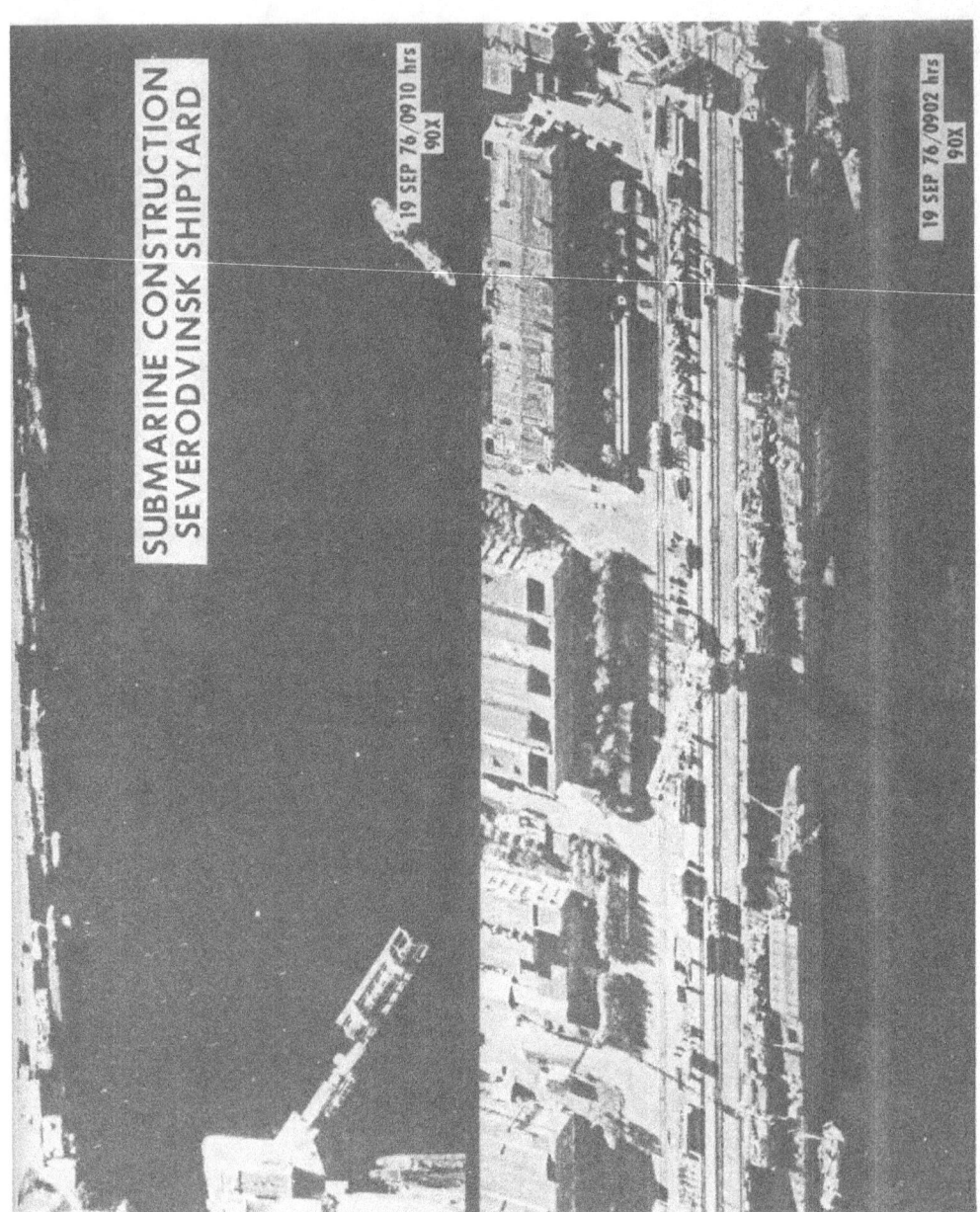

Graphic 47. Submarine Construction, Severodvinsk Shipyard

- **Hexagon Quality**. The high quality of imagery that HEXAGON was capable of achieving is sometimes overlooked because the GAMBIT program completely overlapped HEXAGON and produced imagery of the very highest quality. As pointed out in the text, HEXAGON was capable of meeting a high proportion of the Intelligence Community's surveillance requirements. A number of examples (Graphics 50-60) of high-quality HEXAGON imagery are included to illustrate this point. The two US images provide the reader with familiar objects for comparative purposes.

Graphic 50. New York City World Trade Center – Mission 1213-1 40X

Graphic 51. Shea Stadium – Mission 1216-3 80X

Graphic 52. Soviet Kiev-class Carrier, Black Sea, USSR – Mission 1216-1 80X

Graphic 54. TALL KING Radar – Mission 1217-2 150X

Graphic 55. Simferopol USSR, Deep-Space Tracking Antenna – Mission 1217-3

Graphic 56. Severodvinsk Shipyard, USSR – Typhoon-class Submarine 80X

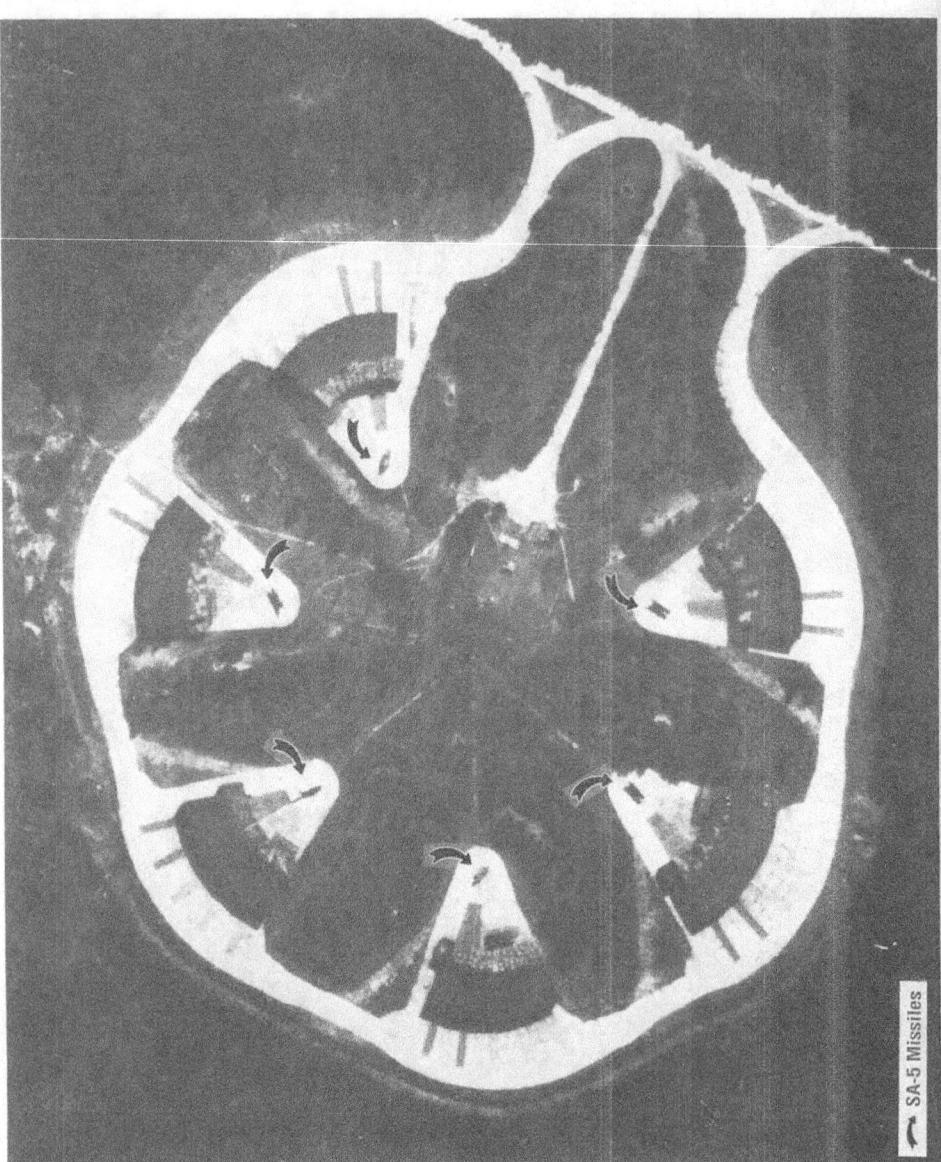

Graphic 57. Soviet SA-5 Missiles on Launchers – Mission 1219-1 80X

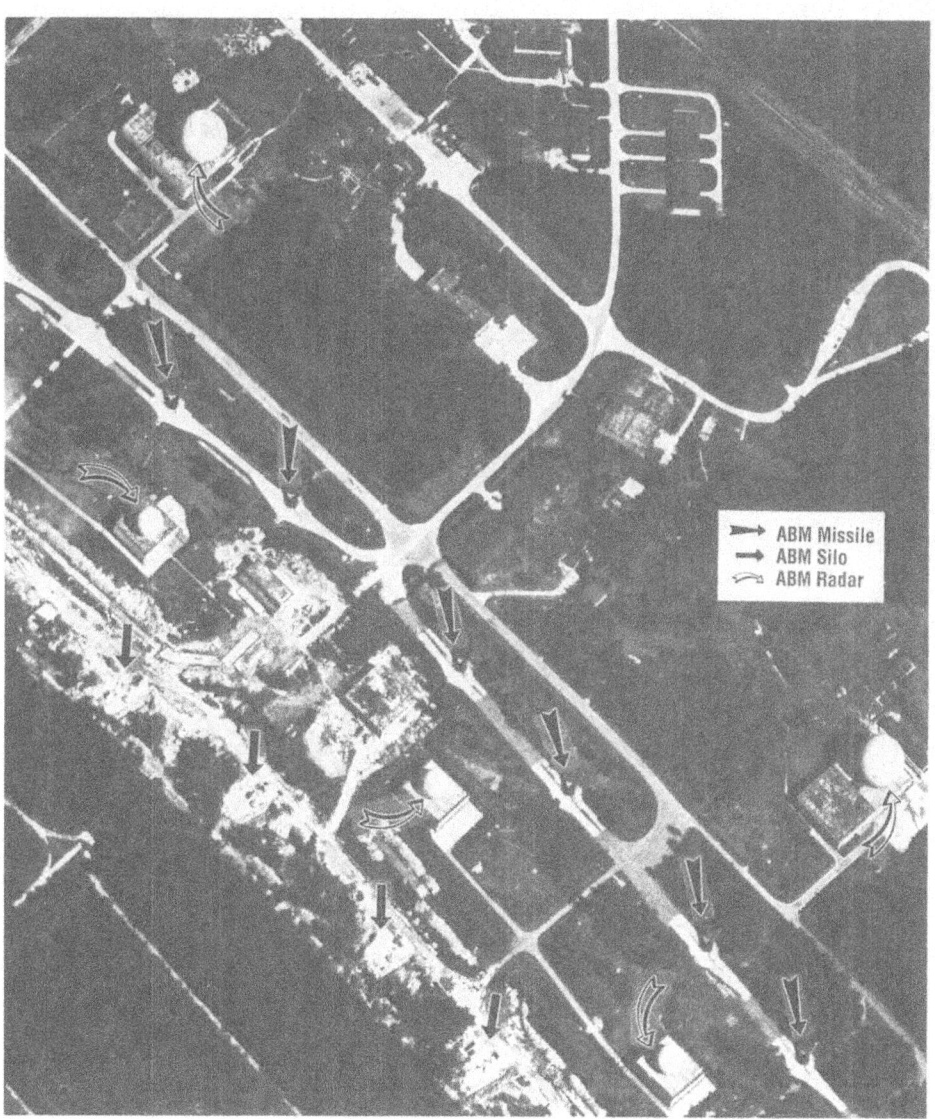

Graphic 58. Moscow ABM Complex – Mission 1218-1 40X

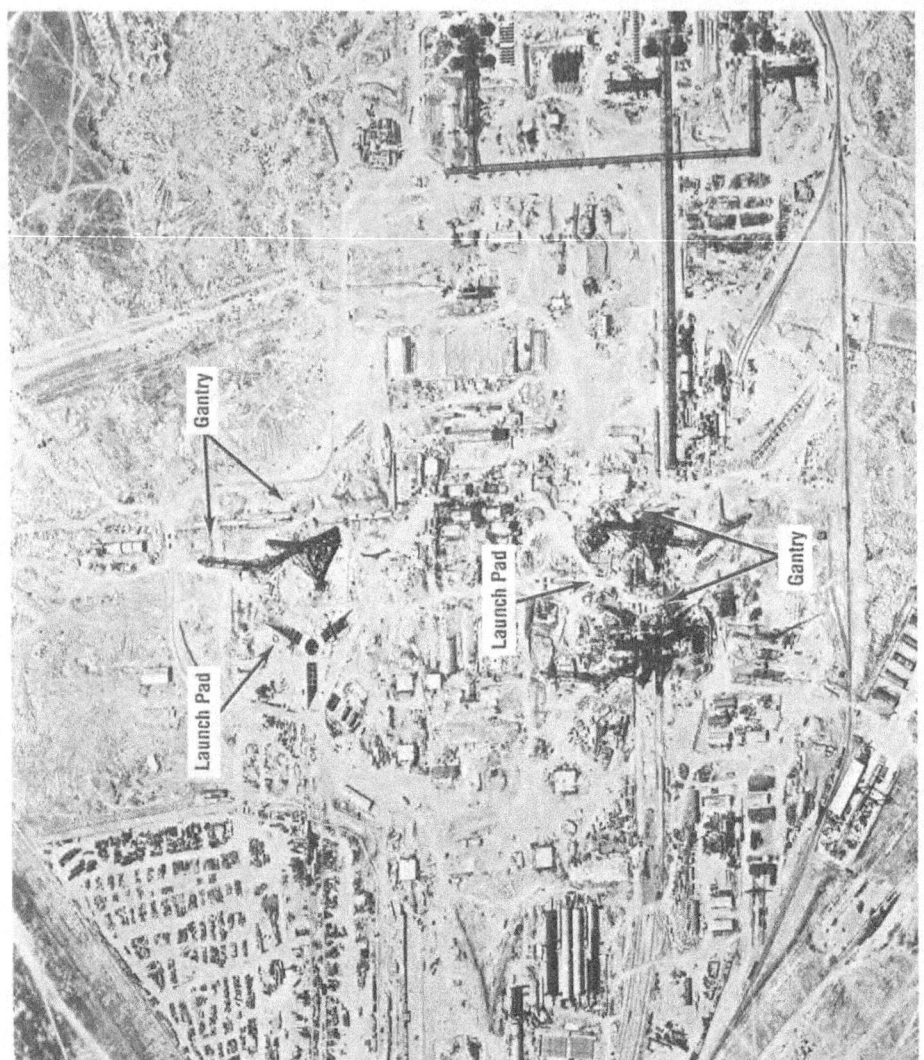

Graphic 59. Tyuratam Missile Test Range, Complex J – Mission 1219-1, 20X

HEXAGON Success — A Team Effort

An attempt has been made in this annex to briefly describe national intelligence management roles and the interfaces between the requirements manager, COMIREX, and the system operator, NRO. In reality these roles are considerably more complicated than can be shown here and they are affected by factors not addressed or merely alluded to but which are part of the overall formula for program success. Some of these factors are:

- development and production of system hardware and software;
- launching of satellites and their maintenance on orbit;
- operations of the film-bucket-recovery force;
- film technology and the development and operation of means for processing and duplicating high-resolution films;
- development of film-exploitation techniques, equipment, and data bases;
- National Tasking Plan for the management of product exploitation;
- distribution of photography and information derived from it;
- development and operation of collection history data bases and graphics displays;
- decompartmentation, sanitization, and decontrol of photography and information derived from it;
- development, review, and management of photography requirements; and
- use of photography for mapping, charting, and geodetic products.

Appendix B

The CORONA Program*

The CORONA Program was approved for development by President Eisenhower on 7 February 1958. At White House direction, the program was organized under the joint leadership of Richard M. Bissell, Jr., CIA, and Brig. Gen. Osmund J. Ritland, US Air Force. CORONA was a breakout from a large Air Force satellite reconnaissance development called WS-117L, which was being conducted at the Air Force Ballistic Missile Division (AFBMD) in Inglewood, California. A portion of WS-117L, called Discoverer, was the precursor of and cover for CORONA. The public was told that Discoverer was for biomedical and other space experiments.

The AFBMD was responsible for all hardware required for CORONA except the payload and, additionally, for providing launching, tracking, and recovery facilities to the program. The CIA funded the camera development, procured the reentry vehicles, provided security supervision for the "black" aspects of the program, and defined its covert objectives.

The Lockheed Missiles and Space Division (under contract to both the CIA and AFBMD) integrated all equipment, developed the upper (spacecraft) stage, and furnished leadership in testing, launching, and on-orbit control operations. Itek developed the camera; General Electric built the recovery capsule; and Douglas furnished the Thor boosters.

CORONA security kept the program "black." This was possible because to the uncleared world CORONA could be presented as Discoverer, a technological program for exploring the space environment and for pioneering assistance to later satellites. The CORONA launching site was at Vandenberg Air Force Base; its control station was at Sunnyvale; and recovery ships and aircraft worked out of Oahu, Hawaii.

CORONA Number 1 was launched on 28 February 1959 purely as a testbird. In a subsequent series of 11 flights, extending to August 1960, there were no complete successes although significant progress was made. Flight number 13, a diagnostic flight carrying only test instrumentation, was recovered by water-pickup on 12 August 1960. But the first actual success—with success measured in terms of exposed film delivered—was flight number 14, which was air-recovered on 18 August 1960.

---

*See also F.E. Oder, James C. Fitzpatrick, Paul E. Worthman. The CORONA Story, December 1988, BYE 140002-88.

~~SECRET~~
~~NOFORN-ORCON~~

In the first two years of operation, dating from 18 August 1960, 48 photographic missions were attempted with 19 true successes. The original camera, known as the KH-1, produced nominal resolutions of 40 feet; with improvement in camera and film, resolutions began to move below 10 feet. There was continual improvement in the CORONA system. A stereoscopic arrangement, called CORONA-M and known as the KH-3, was introduced in 1962. In 1963, CORONA-Js (also called the KH-4), capable of carrying 15,000 feet of film in each of *two* reentry capsules, were launched. The CORONA J-3 (known as the KH-4B), initiated as the CORONA Improvement Program in 1965 and first flown in 1967, obtained ground resolutions of 6 to 8 feet.

CORONA's lifespan, as a program, was 12 years and covered 145 launchings. Ground resolutions of 6 to 10 feet were eventually achieved. By 1970, CORONA could remain in orbit for 19 days, make operational responses to cloud cover, provide accurate mapping information, and return coverages as large as 8,400,000 square nm. The final cost of an average mission was ███████████.

The Intelligence Community described CORONA's contribution to its resources as "virtually immeasurable."

Appendix C

The GAMBIT Program*

With the termination of U-2 flight operations over the Soviet Union in May 1960, it was apparent to President Eisenhower and his senior advisors that satellite photography, the only alternative to aircraft overflight, would require added capability to fill the gap in intelligence data. The photographic satellite, CORONA, which was just getting into operation, could not provide the high resolution needed for detailed target identification. Eisenhower directed his science advisor, Dr. George Kistiakowsky, to gather an advisory group to study this problem. The members of this group recommended a new initiative within the Air Force's extant WS-117L Samos Program.

Coincidentally, Air Force Under Secretary Dr. Joseph V. Charyk had knowledge of an Eastman Kodak Company (EK) suggestion for a system that could get two-to-three-foot ground resolution. The system would use a 77-inch focal-length f/4.0 lens. The suggestion was adopted and resulted in the GAMBIT-1 system, also known as the KH-7. General Electric (GE) was chosen to build the orbital-control vehicle (OCV) and the film-carrying reentry vehicle (RV), which would be put in space by an Atlas/Agena booster system. GAMBIT was managed by Program A (SAFSP).

After a difficult development, the program had a successful first flight in July 1963. A number of flights followed, each characterized by moderate to fatal technical problems in the OCV. After some strenuous Air Force management pressure, remedial action by GE made it possible for 14 of the last 15 flights (of the total of 38 GAMBIT-1 flights) to be rated as very successful. In these latter flights, best resolutions ranged from ▓▓▓ 2.5 feet, with flight durations up to eight days and ▓▓▓▓▓▓▓▓ targets covered per flight. The last GAMBIT-1 KH-7 flight was in June 1966.

The GAMBIT-1 series was replaced by a more optimal GAMBIT-3 series. Designed in 1963 and started as a hardware program in early 1964, GAMBIT-3 had a very successful first flight in mid-1966. GAMBIT-3 consisted of a photographic-payload section made by EK (and shown as the KH-8 camera) and a satellite control section made by Lockheed Missiles and Space Company, Inc. GAMBIT-3 contained a 175-inch f/40 lens, had stereo capability, carried 10,000 feet of 9.5-inch ultra-thin base film with an aerial index of 6.0, and was capable of ▓▓▓▓▓▓▓▓▓▓▓▓▓▓▓▓▓▓▓▓▓▓▓▓. It was launched by a Titan-IIIB booster. It originally flew with one GE RV that was very similar to the proven CORONA RV. Beginning with

---

*See also, F.C.E. Oder, James C. Fitzpatrick, Paul E. Worthman, The GAMBIT Story, 1990, BYE 140002-90.

GAMBIT-3 number 23, the payload was CORONA RV. increased to include two RVs for the remainder of the program; the 54th and last GAMBIT-3 flight took place in April 1986. Of the 54 flights, 51 were quite successful. Three failed to reach orbit as a result of one Titan failure and two Agena failures. On the earliest flights, best resolutions were generally ▓▓▓ inches, improving by the 10th flight to a consistent ▓▓▓, by the 30th to ▓▓▓, by the 41st ▓▓▓ and for the last 10 flights (other than one) a best resolution of ▓▓▓▓▓▓. Flight durations began at about one week; by the end of the program they extended well beyond 100 days.

Appendix D

13 August 1965

Agreement For Reorganization of The National Reconnaissance Program

A. The National Reconnaissance Program

1. The NRP is a single program, national in character, to meet the intelligence needs of the Government under a strong national leadership, for the development, management, control, and operation of all projects, both current and long range for the collection of intelligence and of mapping and geodetic information obtained through overflights (excluding peripheral reconnaissance operations). The potentialities of US technology and all operational resources and facilities must be aggressively and imaginatively exploited to develop and operate systems for the collection of intelligence which are fully responsive to the Government's intelligence needs and objectives.

2. The NRP shall be responsive directly and solely to the intelligence collection requirements and priorities established by the United States Intelligence Board. Targeting requirements and priorities and desired frequency of coverage of both satellite and manned aircraft missions over denied areas shall continue to be the responsibility of USIB, subject to the operational approval of the 303 Committee.

B. The Secretary of Defense will:

1. Establish the NRO as a separate agency of the DoD and will have the ultimate responsibility for the management and operation of the NRO and the NRP;

2. Choose a Director of the NRO who will report to him and be responsive to his instructions;

3. Concur in the choice of the Deputy Director of the NRO who will report to the DNRO and be responsive to his instructions;

4. Review and have the final power to approve the NRP budget;

5. Sit with members of the Executive Committee, when necessary to reach decisions on issues on which committee agreement could not be reached.

C. The Director of Central Intelligence will:

1. Establish the collection priorities and requirements for the targeting of NRP operations and the establishment of their frequency of coverage;

2. Review the results obtained by the NRP and recommend, if appropriate, steps for improving such results;

3. Sit as a member of the Executive Committee;

4. Review and approve the NRP budget each year;

5. Provide security policy guidance to maintain a uniform system in the whole NRP area.

D. National Reconnaissance Program Executive Committee:

1. An NRP Executive Committee, consisting of the Deputy Secretary of Defense, the Director of Central Intelligence, and the Special Assistant to the President for Science and Technology, is hereby established to guide and participate in the formulation of the NRP through the DNRO. (The DNRO will sit with the Executive Committee but will not be a voting member.) If the Executive Committee cannot agree on an issue, the Secretary of Defense will be requested to sit with the Committee in discussing this issue and will arrive at a decision. The NRP Executive Committee will:

    a. Recommend to the Secretary of Defense an appropriate level of effort for the NRP in response to reconnaissance requirements provided by USIB and in the light of technical capabilities and fiscal limitations.

    b. Approve or modify the consolidated NRP and its budget.

    c. Approve the allocation of responsibility and the corresponding funds for research and exploratory development for new systems. Funds shall be adequate to ensure that a vigorous research and exploratory development effort is achieved and maintained by the DoD and CIA to design and construct new sensors to meet intelligence requirements aimed at the acquisition of intelligence data. This effort shall be carried out by both CIA and DoD.

d. Approve the allocation of development responsibilities and the corresponding funds for specific reconnaissance programs with a view to ensuring that the development, testing, and production of new systems is accomplished with maximum efficiency by the component of the Government best equipped with facilities, experience, and technical competence to undertake the assignment. It will also establish guidelines for collaboration between departments for the mutual support where appropriate. Assignment of responsibility for engineering development of sensor subsystems will be made to either the CIA or DoD components in accordance with the above criteria. The engineering development of all other subsystems, including spacecraft, reentry vehicles, boosters, and booster interface subsystems, shall in general be assigned to an Air Force component, recognizing, however, that sensors, spacecraft, and reentry vehicles are integral components of the system, the development of which must proceed on a fully coordinated basis with a view to ensuring optimum system development in support of intelligence requirements for overhead reconnaissance. To optimize the primary objective of systems development, design requirement of the sensors will be given priority in their integration within the spacecraft and reentry vehicles.

e. Assign operational responsibility for various types of manned overflight missions to CIA or DoD subject to the concurrence of the 303 Committee.

f. Periodically review the essential features of the major program elements of the NRP.

2. The Executive Committee shall meet on the call of either the Deputy Secretary of Defense or the Director of Central Intelligence. All meetings will be attended by the DNRO and such staff advisors as the Deputy Secretary of Defense or the Director of Central Intelligence consider desirable.

E. National Reconnaissance Office

1. To implement the NRP, the Secretary of Defense will establish the NRO as a separate operating agency of the DoD. It shall include the Satellite Operations Center (SOC) which shall be jointly manned.

2. The DNRO shall be appointed by the Secretary of Defense. The DNRO will:

a. Subject to direction and control of the Secretary of Defense and the guidance of the Executive Committee as set forth in Section D above, have the responsibility for managing the NRO and executing the NRP.

b. Subject to review by the Executive Committee, and the provisions of Section D above, have authority to initiate, approve, modify, redirect, or terminate all research and development programs in the NRP. Ensure, through appropriate recommendations to the Executive Committee for the assignment of research and development responsibilities and the allocation of funds, that the full potentialities of agencies of the Government concerned with reconnaissance are realized for the invention, improvement, and development of reconnaissance systems to meet USIA requirements.

c. Have authority to require that he be kept fully and completely informed by all agencies and departments of the Government of all programs and activities undertaken as part of the NRP.

d. Maintain and provide to the members of the Executive Committee records of the status of all projects, programs, and activities of the NRP in the research, development, production, and/or operational phases.

e. Prepare a comprehensive budget for all aspects of the NRP.

f. Establish a fiscal control and accounting procedure to ensure that all funds expended in support of the NRP are fully accounted for and appropriately utilized by the agencies concerned. In particular, the budget shall show separately those funds to be applied to research and exploratory design development, systems development, procurement, and operational activities. Funds expended or obligated under the authority of the Director of Central Intelligence under Public Law 110 shall be administered and accounted for by CIA and will be reported to DNRO in accordance with agreed-upon procedures.

g. Sit with the USIB for the matters affecting the NRP.

3. The Deputy Director of NRO shall be appointed by the Director of Central Intelligence with the concurrence of the Deputy Secretary of Defense and shall serve full time in a line position directly under the DNRO. The Deputy Chief shall act for and exercise the powers of the DNRO during his absence or disability.

4. The NRO shall be jointly staffed in such a fashion as to reflect the best talent appropriately available from the CIA, the three military departments, and other Government agencies. The NRO staff will report to the DNRO and Deputy Director of NRO and will maintain no allegiance to the originating agency or department.

F. Initial Allocation of Program Responsibilities

1. Responsibility for existing programs of the NRP shall be allocated as indicated in Annex A attached hereto.

(Signed) Cyrus Vance	(Signed) W.F. Raborn
Deputy Secretary of Defense	Director of Central Intelligence

13 August 1965

Annex A

The following assignments for the development of new optical sensor subsystems are made to take full advantage of technical capability and experience of the agencies involved.

1. The CIA will develop the improvements in the CORONA general search optical sensor subsystems.

2. Following the selection of a concept and a contractor for full-scale development in the area of advanced general search, the CIA will develop the optical sensor subsystem for that system.

3. The Air Force (SAFSP) will develop the G-3 optical sensor subsystem for the advanced high-resolution pointing system.

4. SAFSP will develop the optical sensor subsystems (manned and unmanned) for the MOL program.

The DNRO will, in managing the corresponding overall system developments, ensure that:

1. The management of and contracting for the sensors is arranged so that the design and engineering capabilities in the various contractors are most efficiently utilized.

2. The sensor packages and other subsystems are integrated in an overall system engineering design for each system, with DNRO having responsibilities for systems integration of each overall system.

SECRET
NOFORN-ORCON

Appendix E

HEXAGON and the Space Transportation System

In 1973, the Satellite Basic Assembly (SBA) contractor Lockheed Missiles and Space Company (LMSC) and the camera systems contractor Perkin-Elmer (PE) were tasked by SAFSP to study a HEXAGON satellite vehicle designed specifically for use with the space transportation system (STS)—also commonly known as the space shuttle—then under development by NASA. The contractors were to formulate operational and design concepts and estimate system cost. Three operational concepts were considered individually and in combination:

- Resupply: on-orbit replacement of expendables

- Maintenance: on-orbit replacement of failed or life-limited items

- Refurbishment: return to earth and restoration to flight configuration

A prior study of compatibility of the HEXAGON satellite vehicle (SV) with the STS had been completed in January 1972. The objectives of that study had been to develop and describe the minimum modifications required to make the SV and its supporting Aerospace Ground Equipment (AGE) and facilities compatible with the STS and to estimate incremental costs associated with such modifications. Two primary modes of SV/STS operation were considered: booster substitution, in which the STS would be used only as a booster; and boost/retrieval, in which the STS would be used as an SV booster and a retrieval vehicle, with refurbishment and reuse of the SV after retrieval.[278] The 1973 study was, therefore, a follow-on effort aimed at examining the extent of SV design change to more fully use the capability of the STS.

The HEXAGON Block-III SV was used as the point of departure for these studies. General study criteria [239] were:

- Two missions to be conducted per year, each for a minimum of 120 days, with the first HEXAGON SV/STS launching in 1982 from Vandenberg AFB. SV/STS return from orbit was assumed to take place at VAFB. A 10-year operational program was priced "with provision made for a continuing program beyond the 1992 cut-off for pricing."

- The size, type and quantity of RF's were variables, as was the frequency of data return. All film was to be returned by RV's "except that the last portion of the mission could be retained on board the SV and returned by the STS during and SV retrieval or resupply mission."

- Ground control and monitoring of the STS and SV during on-orbit operations was to be done by the Air Force Satellite Control Facility.

SECRET
Handle via
BYEMAN-TALENT-KEYHOLE
Control Systems Jointly
BYE 140003-92

- The reliability goal of the SV for STS operations would be the same as for Block-III SVs, namely 0.85 for 60 days (excluding camera and RV-separation systems), with SV deployment/retrieval operations having "a higher reliability goal."

- SVs would retain a deboost capability, so that, in the event an SV could not be retrieved by the STS, it could be deorbited into a deep ocean area.

- "All vehicles were assumed to be launched into the same basic sun-synchronous orbit currently employed by the HEXAGON program: 96.4 degrees inclination with the argument of perigee being located at 45 degrees North latitude." (This would necessitate the space-shuttle overflying the Sino-Soviet landmass.)

A space replacement unit (SRU) approach was assumed in the resupply and maintenance study. Fluid and pressurant transfer was also considered. An important factor in the operational concept was the fact that "approximately half of the HEXAGON vehicle weight is in expendables (fuel, film, RVs, and so forth)." After considering the on-orbit resupply/maintenance modes of formation flying (no physical coupling between SV and STS), soft dock (SV/STS spatial orientation provided by a minimum of one remote manipulator arm), and hard-dock (SV rigidly attached to the STS), it was concluded that the SV should be hard-docked to the STS and exchange of SRUs would be accomplished by program-provided special equipment. It should be noted, at this juncture, that later LMSC experience on NASA's Hubble Space Telescope, which has SRUs, showed that because of the EVA-suited astronaut's physical limitations, significant design constraints are placed on SRUs, particularly on those areas where precise location/orientation, complex electrical connections (use of multi-pin connectors), and fluid connections are involved. Whether or not the precision required in the orientation and location of the film-path through the replacement RVs could be achieved was not demonstrated during the study.

Trade studies were performed on various candidate vehicles; for example, trading number of RVs against propellant load. "Conceptually all identified configurations seemed technically feasible. Cost was the most significant variable among configuration . . . The primary cost-driver is the non-recurring cost for development of resupply kits, special STS-mounted equipment, and configuring the SV for resupply/maintenance . . . Therefore, a non-resupply operational concept was selected."[280] In other words, the concept was to return the SV after use, refurbish it on earth, and then return the refurbished SV to orbit.

Although the concept of reusing space hardware was attractive and technically feasible, the very significant non-recurring costs associated with this proposed approach led to its demise in the early 1970s.

The concept of reusability for the HEXAGON spacecraft did not end with the 1973 study. In 1982, the idea was revived by Maj. Gen. John Kulpa, director of SAFSP, and studies were begun with contractors. Kulpa arranged for NASA people, including some of the astronauts, to be cleared for HEXAGON. Instead of a major rebuild of the SV, as the 1973 study had envisaged, Kulpa's idea was to allow only minimal essential changes for stowing the HEXAGON vehicle in the shuttle bay and for accommodating a different (dynamic) launching environment. He hoped to launch the last two HEXAGON vehicles (19 and 20) by STS from Kennedy Space Center. His plans did not materialized for a variety of reasons—mostly cost—but also including NASA's understandable reluctance to launch on a northbound trajectory across the eastern United States with solid-rocket motor separation occurring near Cleveland, Ohio.

Appendix F

Key Personnel on the HEXAGON Program

GOVERNMENT PERSONNEL

Director, National Reconnaissance Office

Dr. Brockway McMillan	Mar 1963 to Oct 1965
Dr. Alexander H. Flax	Oct 1965 to Mar 1969
Dr. John L. McLucas	Mar 1969 to Dec 1973
Mr. James W. Plummer	Dec 1973 to Jun 1976
Dr. Charles W. Cook (Acting)	Jun 1976 to Aug 1976
Mr. Thomas C. Reed	Aug 1976 to Apr 1977
Dr. Charles W. Cook (Acting)	Apr 1977 to Aug 1977
Dr. Hans Mark	Aug 1977 to Oct 1979
Dr. Robert J. Hermann	Oct 1979 to Aug 1981
Mr. Edward C. Aldridge, Jr.	Aug 1981 to Apr 1986[281]

Director, SAFSP

Brig. Gen. John L. Martin	Jul 1965 to Jul 1969
Brig. Gen. William G. King	Aug 1969 to Mar 1971
Brig. Gen. Lew Allen, Jr.	Apr 1971 to Jan 1973
Brig. Gen. David D. Bradburn	Jan 1973 to Jul 1975
Brig. Gen./Maj Gen. John E. Kulpa, Jr.	Aug 1975 to Jan 1983
Brig. Gen./Maj Gen. Ralph H. Jacobson	Jan 1983 to Apr 1986

Director, Office of Special Projects (OSP) CIA

John J. Crowley	Sep 1965 to Nov 1970
Harold L. Brownman	Nov 1970 to Jun 1973[282]

HEXAGON System Program Office (SPO) SAFSP

Col. Frank S. Buzard	Jul 1966 to Jun 1971
Col. Robert H. Krumpe	Jun 1971 to Aug 1973
Col. Raymond A. Anderson	Aug 1973 to Jul 1978
Col. Lester S. McChristian	Jul 1978 to Mar 1983
Col. Larry Cress	Mar 1983 to Apr 1986

HEXAGON Sensor Subsystem Program Office (SSPO) CIA-OSP

[REDACTED]	Sep 1965 to Jun 1973[282]

CONTRACTOR PERSONNEL

Lockheed Missiles and Space Company

Satellite Basic Assembly & System Integration

Dr. Stanley I. Weiss	Jul 1967 to Feb 1970
Paul J. Heran	Feb 1970 to Feb 1980
Steve P. Treat	Feb 1980 to 1983
Bob Johnson	1983 to 1986

McDonnell/Douglas Astronautics Company

Mark 8 Reentry Vehicle

Logan T. MacMillan	Jun 1968 to 1974
Forrest D. Blanton	1974 to 1984
Fred Goetsch	1984 to 1986

General Electric Company, Aerospace Electronics Systems Department

Extended Command System

John H. Griswald	1964 to 1966
Norman N. Feldman	1966 to 1967
Robert M. Larkin	1967 to 1969
James O. Moore	1969 to 1978
Elmer B. Tamanini	1978 to 1980
Francis Smith	1980 to 1986

Thompson-Ramo-Wooldridge Corporation

T'Unity Software

Thomas A. Magness	Oct 1969 to Nov 1971
Winston W. Royce	Nov 1971 to Mar 1972
William V. Buck	Mar 1972 to Sep 1972
Gerald K. Lambert	Sep 1972 to Nov 1973
David M. Yaksick	Nov 1973 to Apr 1975
Clair D. Calvin	Apr 1975 to Aug 1979
David M. Yaksick	Aug 1979 to 1986

Itek Corporation Optical Systems Division

Mapping Camera Module

John T. Watson	Jun 1968 to Nov 1968
John F. Doyle	Nov 1968 to Jan 1970
Paul J. Mailhot	Jan 1970 to Feb 1972
D. David Cook	Feb 1972 to Aug 1973
William J. Reusch	Aug 1973 to Apr 1974
Jean R. Manent	Apr 1974 to Oct 1976
Maurice G. Burnett	Oct 1976 to program completion 1981

General Electric Company, Reentry Systems Division

Mark V Reentry Vehicle

Stephen F. Csencsitz	Mar 1970 to Feb 1975
John S. Kleban	Feb 1975 to program completion 1981

Aerospace Corporation

General Systems Engineering

George M. Kelsey	Jul 1966 to Dec 1966
John D. Sorrels	Dec 1966 to Jul 1967
John W. Luecht	Aug 1967 to Dec 1968
Leonard C. Lidstrom	Jan 1969 to Aug 1969
Bruce L. Adams	Dec 1969 to Apr 1973
C. James Crickmay	Apr 1973 to Jun 1979
James R. Henry	Jun 1979 to Mar 1983
Bert Larkin	Mar 1983 to Mar 1986

Perkin-Elmer Corporation

Sensor Subsystem

Michael F. Maguire	Oct 1966 to Aug 1969
Bernard Malin	Aug 1969 to Jan 1971
Paul E. Petty	Jan 1971 to Apr 1973
B. Alan Ross	Apr 1973 to Sep 1975
Bernard Malin	Sep 1975 to Dec 1977
Michael A. Mazaika	Dec 1977 to Jul 1979
Kent H. Meserve	Jul 1979 to Oct 1980
Vic Abramson	Oct 1980 to Jan 1985
Leonard J. Farkas	Jan 1985 to Apr 1986

Glossary Of Acronyms

ABM	Antiballistic Missile
ABMA	Army Ballistic Missile Agency
ACS	Attitude-Control System
AEC	Atomic Energy Commission
AFBMD	Air Force Ballistic Missile Division
AFSC	Air Force System Command
ARDC	Air Research and Development Command
ARM	Attitude-Reference Module
ARPA	Advanced Research Projects Agency
BAS	Broad-Area Search
BRAC	Back-up Recovery Attitude
BYE	Byeman
CAC	Civil Applications Committee
CAMS	Comirex Automated Management System
CCD	Charge-coupled device
CC&D	Camouflage, Concealment, and Deception
CDR	Critical Design Review
CIA	Central Intelligence Agency
CIA-OSP	CIA Office of Special Projects
COMIREX	Committee on Imagery Requirements and Exploitation
COMOR	Committee for Overhead Reconnaissance
CORN	Controlled Optical-Range Network
DCI	Director of Central Intelligence
DDCI	Deputy Director of Central Intelligence
DDR&E	Director of Defense Research and Engineering
DDS&T	CIA Deputy Director for Science and Technology
DIA	Defense Intelligence Agency
DMA	Defense Mapping Agency
DNRO	Director, National Reconnaissance Office
DoD	Department of Defense
DSPD	Deputy System Program Director
ECS	Extended Command System
EK	Eastman Kodak
ELINT	Electronic Intelligence
EOI	Electro-optical imaging
EPM	Electrical Power Module
ExCom	NRO Executive Committee
EXSUBCOM	Exploitation Subcommittee
GE	General Electric
GWC	Global Weather Central
IC	Intelligence Community
ICBM	Intercontinental Ballistic Missile
ICRS	Imagery Collection Requirements Subcommittee
IFWG	Interface Working Group
IR	Infrared
IRBM	Intermediate-Range Ballistic Missile

LMSC	Lockheed Missile and Space Company
LOC	Line of Communication
MER	Manned Earth Reconnaissance
MER-1	Manned Earth-Reconnaissance System (U.S. Navy proposal)
MC&G	Mapping, Charting, and Geodesy
MCS	Mapping Camera System
MCS	Minimum Command System
MIDAS	
MIT	Massachusetts Institute of Technology
NAA	North America Aviation
NACA	National Advisory Committee on Aeronautics
NASA	National Aeronautics and Space Administration
NATO	North Atlantic Treaty Organization
NIIRS	National Imagery Interpretability Rating Scale
NPIC	National Photographic Interpretation Center
NRL	Naval Research Laboratory
NRO	National Reconnaissance Office
NRP	National Reconnaissance Program
NRT	Near-real-time
NSC	National Security Council
NTP	National Tasking Plan
NVR	Non-Volatile Residue
OAM	Orbit-Adjust Module
OAS	Orbit-Adjust System
OCV	Orbital-Control Vehicle
OD-4	Operating Division-4/SAFSS
ONR	Office of Naval Research
OTD	Optical Technology Division (P-E)
PACS	Primary Attitude-Control System
PDR	Preliminary Design Review
P-E	Perkin-Elmer
PET	Performance Evaluation Team
PFIAB	President's Foreign Intelligence Advisory Board
PMR	Pacific Missile Range
PMU	Programmable Memory Unit
PPMU	Parallel Programmable Memory Units
PPS	Photographic Payload Section (GAMBIT Vehicle)
PSAC	President's Science Advisory Committee
QPR	Quarterly Program Review
R&D	Research and development
RACS	Redundant Attitude-Control System
RAND	Research and Development Corporation
RCA	Radio Corporation of America
RCM	Reaction-Control Module
RCS	Reaction-Control Subsystem
RCT	Reaction-Control Thrusters
RFP	Request for Proposal
RTS	Remote Tracking Station
RV	Reentry Vehicle

R-W	Ramo-Wooldridge Corporation
S³	Solid-State Sensor (Stellar Camera System)
SAFMS	Secretary of the Air Force/Missiles and Space (Office)
SAFSP	Secretary of the Air Force/Special Projects (Office)
SAFSS	Secretary of Air Force/Space Systems
SALT	Strategic Arms Limitation Talks
SAM	Solar-Array Module
SAS	System Analysis Staff (CIA)
SBA	Satellite Basic Assembly
SCAD	S³ Activity Detector
SCF	Satellite-Control Facility
SDV	Satellite Development Vehicle
SE/TD	System Engineering/Technical Direction
SETS	System Engineering and Technical Support
SI	Stellar-Index Camera
SIGINT	Signal Intelligence
SOC	Satellite Operations Center
SOR	System Operational Requirement
SPD	System Program Director
SPO	System Program Office
SPS	Special Projects Staff (CIA)
SRM	Solid Rocket Motor
SRU	Space Replacement Unit
SS	Sensor Subsystem
SSD	Space Systems Division
SSPO	Sensor Subsystem, Project Office
STC	Satellite Test Center
STL	Space Technology Laboratories
STS	Space Transportation System
SV	Satellite Vehicle
TCP	Technological Capabilities Panel
TFX	Tactical Fighter (Advanced) Experimental
TK	Talent-Keyhole
TRW	Thompson-Ramo-Woolridge Corporation
TTCM	Tracking Telemetry and Command Module
USIB	US Intelligence Board
UTB	Ultra-Thin-Base Film
UUTB	Ultra-Ultra-Thin-Base Film
VAFB	Vandenberg Air Force Base
WAC	World Aeronautical Cell
WADC	Wright Air Development Center
WAG	World Aeronautical Grid
WDD	Western Development Division

Reverse side blank

SECRET
NOFORN-ORCON

Volumn III

Index

AAA

A-12 aircraft 23
Able-Star upper stage 16
Abramson, Victor 243
Adams, Bruce L. 243
Advanced Research Project Agency (ARPA) 5, 12, 14, 17
Aerobee, Aerobee-Hi 8
Aerospace Corporation 38, 49, 74, 86, 96, 101, 121, 243
Aft-section 81
Agena 13, 16, 229
Air Force Ballistic Missile Division (AFBMD) 13, 14, 16, 17, 227
Air Force Systems Command (AFSC) 111
Air Research and Development Command (ARDC) 9, 10, 17
Aldridge, Edward C., Jr. 241
Alexander, Charles C. 255
Allen, Lew, Jr., Maj. Gen. (USAF) 96, 106, 109, 241
American Rocket Society 11
Anderson, Raymond A., Col. (USAF) 107, 112, 241
Antiballistic Missile (ABM) Treaty 120
Apollo moon program 12
Army Ballistic Missile Agency (ABMA) 7, 12
Atlas 2, 9, 229
Attitude reference module (ARM) 81
AVCO 38, 41, 67

BBB

B-52 aircraft 1
Backup Recovery Attitude Control System (BRAC) 82
Baker, James 41
Battle, Lee, Col. (USAF) 14
Bell Laboratories 27, 48
Berganini, David F., Col. (USAF, retired) viii
Bison aircraft 1
Bissell, Richard M., Jr. 3, 14, 19, 20, 22, 24, 47, 227
Blanton, Forrest D. 242
Bradburn, David D., Brig. Gen. (USAF) 106, 110, 111, 241
Bross, John 43
Brownman, Harold L. 109, 241
Brugioni, Dino A. viii
Buck, William V. 242
Bureau of Budget (BoB) 50, 93
Burnett, Maurice G., Col. (USAF, retired) vii, 24
Buzard, Frank S., Col. (USAF) viii, 65, 67, 75, 96, 101, 241
Byeman security system (BYE) 124, 130

CCC

Cabell, Charles P., Lt. Gen. (USAF) 20
California Institute of Technology 3
Calvin, Clair D. 242
Camera A 83, 85, 114
Camera B 83, 85
Campbell, Edward 41
Carter, David L. 41, 56
Carter, Marshall S., Jr., Lt. Gen. (USA) 30, 36, 41
Central Intelligence Agency (CIA) 3, 127
Charyk, Joseph V. 19, 20, 22, 24, 47, 124, 229
██████████ viii
Civil Applications Committee (CAC) 127
Comirex Automated Management System (CAMS) 129, 134
Committee for Overhead Reconnaissance (COMOR) 33, 127
Committee on Imagery Requirements and Exploitation (COMIREX) viii, 127
Comsat Corporation 24
Controlled Optical Range Network (CORN) 101
Cook, Charles W. 241
Cook, D. David 242
Cornell Aeronautical Laboratories 61
CORONA vii, 13, 14, 15, 16, 17, 19, 22, 25, 27, 29, 33, 48, 75, 93, 123, 125, 145, 227, 235
Courier (Army Communication Satellite Program) 12, 13
Cress, Larry, Col. (USAF)114, 241
Crickmay, C. James 243
Crowley, John J. 38, 41, 55, 241
Csencsitz, Stephen F. 243
Cunningham, James A. 49
Curtin, Richard, Brig. Gen. (USAF) 19

DDD

DeLauer, Richard 38
Defense Intelligence Agency (DIA) 50, 127
Defense Mapping Agency (DMA) 74, 111, 127, 133
Deputy Director (of CIA) for Science Technology (DDS&T) 27, 32, 109
Director, Defense Research and Engineering (DDR&E) 30
Director, National Reconnaissance Office (DNRO) 20, 29, 50, 65, 93, 109, 231,241
Dirks, Leslie C. 33, 41, 56, 110
Discoverer Program 13, 14, 227
DoD Research and Development Board 8

SECRET
Handle via
BYEMAN-TALENT-KEYHOLE
Control Systems Jointly
BYE 140003-92

Donovan, Allen F. 31
Donovan Review Committee 86
■■■■■
Douglas Aircraft Corporation 9, 227
Douglas, James H., Jr. 19
Doyle, John F. 242
Drell Committee 31
Drell, Sidney D. 31, 41
DuBridge, Lee 3
Duckett, Carl E. 109
Dulles, Allen W. 14, 20, 27

EEE

Eastman Kodak Co. 39, 44, 88, 97, 229
Eisenhower, Dwight D. 1, 11, 14, 15, 27, 123, 125, 227, 229
Electrical Power Module (EPM) 82
Electro-optical-imaging (EOI) system 120
Epple, Henry 101
Executive Committee (ExCom), NRO 41, 53, 63, 231
Exploitation Subcommittee (EXSUBCOM) 128, 136

FFF

Fairchild Camera & Instrument Company 39, 41, 67
Farkas, Leonard J. 243
Feldman, Norman N. 242
■■■■■
Flax, Alexander H. vii, 52, 55, 64, 241
Forward section 81
Fubini, Eugene G. 30, 34, 39, 41, 49
FULCRUM 27, 29, 30, 31, 32, 33, 34, 39, 40, 68

GGG

GAMBIT vii, 29, 33, 38, 50, 75, 123, 229, 235
Garvin, Richard L. 31
Geary, Leo P., Col. (USAF) 23, 24
General Electric Company 5, 38, 39, 41, 67, 90, 121, 227, 229, 242
Gillette Procedures 14
Gilpatric, Roswell 20, 22, 124
Global Weather Central (GWC) 140
Goetsch, Fred 242
Goodposter, Andrew J. Maj. Gen. (USA) 125
Goodyear Aerospace Corporation 90, 101
Gorman, Frank, Capt. (USN) 56
Greer, Robert E., Maj. Gen. (USAF) 17, 19, 23, 24, 30
Grimwood, James M.
Griswald, John H. 242

HHH

Haas, Donald L. 109
Hard, Donald G., Brig. Gen. (USAF) viii
Harmon, Gary R., Lt. Col. viii

HELIX 63
Helms, Richard M. 20, 46, 55
Henry, James R. 243
Heran, Paul J., Col. (USAF) 39, 56, 242
Hermann, Robert J.
Hill, Jimmie D. viii
Holady, William M. 11
Hornig, Richard 64

III

Imagery Collection Requirements Subcommittee (ICRS) viii, 128
Intelligence Community (IC) 93, 123, 127
Intercontinental ballistic missile (ICBM) 2, 9, 13, 120, 174
Intermediate-range ballistic missile (IRBM) 7, 13, 123
International Geophysical Year (1957) 8, 11
Irving Air Chute Company 101
Itek 14, 29, 39, 41, 67, 75, 121, 242

JJJ

Jacobson, Ralph H., Maj. Gen. (USAF) 114, 241
Jet Propulsion Laboratory (California Institute of Technology) 12
Johns Hopkins University 8
Johnson, Bob 242
Johnson, Lyndon B. 11, 61
Johnson, Roy W. 5
Johnson, V. Alexis 124
Joint Chiefs of Staff (JCS) 5
■■■■■ 75
Jupiter 7

KKK

Keeny, Spurgeon W., Jr. 31
Kelsey, George M. 243
Kennedy, John F. 16, 20, 124
■■■■■
■■■■■ viii
KH-4 65, 131, 228
KH-7 65, 131, 229
KH-8 132, 229
KH-9 131
Kiefer, Eugene 39
Killian, James R., Jr. 3, 4, 10, 11, 13, 14, 27, 41
King, William G., Brig. Gen. (USAF) 96, 241
Kistiakowsky, George B. 15, 19, 20, 22, 229
Kleban, John S. 243
■■■■■ vii
■■■■■ viii
Krumpe, Robert H., Col. (USAF) 101, 107, 109, 241
Kulpa, John E., Jr., Maj. Gen. (USAF) 111, 114, 239, 241

LLL

Lambert, Gerald K. 242
Land, Edwin H. 3, 27, 31, 40, 41, 229
Land Panel 40, 42, 45
Larkin, Bert 243
Larkin, Robert M. 242
Larned, Robert E., Col. (USAF) viii
Levison, Walter vii, 41
Lidstrom, Leonard C. 243
Lifeboat II 111, 116
Lindsay, Richard 41, 42
Ling, Donald P. 31, 41
Livermore Laboratories 27
Lockheed Corporation 9
Lockheed Missiles and Space Company (LMSC) 39, 67, 76, 81, 88, 90, 94, 121, 227, 229, 237, 242
Lowrance, V. L., Adm. (USN) 24
Luecht, John W. 243
Lundahl, Arthur C. 31, 40, 136

MMM

MacLeish, Kenneth 44
MacMillan, Logan T. 242
Madden, Frank 41
Magness, Thomas A. 242
Maguire, Michael 68, 243
Mailhot, Paul J. 242
Malin, Bernard 243
Manent, Jean R. 242
Manned Earth Reconnaissance (MER-1) 8
Mapping camera module (MCM) 86, 106, 114
Mapping, charting, and geodesy (MC&G) 119, 133, 164
Mark, Hans 241
Marsh, Rodger viii
Martin, John L., Jr., Maj. Gen. (USAF) vii, viii, 23, 58, 86, 241
Martin Marietta Corporation 9, 38, 121
Massachussetts Institute of Technology 3
Maxey, Jackson D. 33, 38, 55
Mazaika, Michael A. 243
McChristian, Lester S., Col. (USAF) 112, 114, 241
McCone, John A. 20, 25, 27, 30, 32, 34, 41
McDonnell Aircraft Corporation 67, 74
McDonnell-Douglas Corp. 85, 90, 101, 242
McElroy, Neil 5
McLucas, John L. 93, 109, 241
McMahon, John N. vii, 33, 42, 56
McMillan, Brockway 27, 29, 30, 34, 39, 52, 55, 241
McNamara, Robert S. 20, 22, 48, 55, 61
Medaris, John B., Maj. Gen. (USA) 7
Meinel, Aden B. 31
Meserve, Kent H. 243
Midas program 13, 16
Midsection 81
Missile gap 2
Mission 1201 95
Mission 1202 102
Mission 1203 103
Mission 1204 105
Mission 1205 106
Mission 1206 106
Mission 1207 110
Mission 1208 110
Mission 1209 110
Mission 1210 110
Mission 1211 110
Mission 1212 110
Mission 1213 111
Mission 1214 112
Mission 1215 112
Mission 1216 113
Mission 1217 114
Mission 1218 114
Mission 1219 116
Mission 1220 116
Moore, James O. 242
Morser, Cal 41

NNN

Naka, Robert 93
National Academy of Sciences 11
National Advisory Committee on Aeronautics (NACA) 5, 11, 12
National Aeronautics and Space Act (1958) 11
National Aeronautics and Space Administration (NASA) 5, 12, 13, 16, 237
National Imagery Interpretability Rating Scale (NIIRS) 137
National Photographic Interpretation Center (NPIC) viii, 29, 127, 136
National Reconnaissance Office (NRO) vii, 19, 20, 34, 50, 123, 127, 233
National Reconnaissance Program (NRP) vii, 231
National Security Council (NSC) 3, 15
National Security Agency (NSA) 50, 127
NSAM 156 Committee 124
National Tasking Plan (NTP) 225, 246
Naval Bureau of Aeronautics 8
Naval Research Laboratory (NRL) 8
Near-real-time (NRT) 120
Nimitz, Chester 44
Nitze, Paul 74
Nixon, Richard M. 93, 109
Norman, L. S., Col (USAF) 93
North American Aviation Corporation 67
North Atlantic Treaty Organization (NATO) 1

OOO

Oder, F.C.E. viii
Office of Defense Mobilization Science Advisory Committee 3

Office of Naval Research (ONR) 8
Office of Special Projects (OSP), CIA 55, 61, 109, 241
Oldham, Max S. 40
Operating Division (OD-4) 129
Optical bar 42, 83
Optical Technology Division (OTD), Perkin-Elmer Company 68
Orbit adjust module (OAM) 82

PPP

Pacific Missile Range (PMR) 32
Para Dynamics Inc. 101
▬▬▬▬▬ viii
▬▬▬▬▬ vii, 65, 67, 109, 241
Perkin-Elmer Company 39, 41, 44, 67, 70, 81, 88, 90, 93, 94, 109, 110, 121, 237, 243
Perry, Robert vii
Petty, Paul E. 243
Philbrick, Richard 41
▬▬▬▬▬ 93
Plummer, James W. 241
Polaroid Corporation 3, 43
POPPY 23, 24
President's Foreign Intelligence Advisory Board (PFIAB) 46
President's Science Advisory Committee (PSAC) 11
Program 417 140
Program A vii, 23, 25, 50, 59, 109, 111, 114, 229
Program B vii, 23, 25, 32, 50, 109
Program C 23, 25
Program D 24, 25
Project FEEDBACK 9
Project Paperclip 7
Project Pied Piper 9
Puckett, Allen 41
Purcell Committee 27, 31
Purcell, Edwin 27, 41

RRR

RCA Corporation 9
Raborn, William F. 46, 53, 64, 235
Rand Corporation 3, 9, 123
Rauth, Donald
Reaction control module (RCM) 81, 103
Reconnaissance panel (PSAC) 54
Redstone missile 7
Reed, Thomas C. 241
Reentry vehicle/recovery vehicle (RV) 65, 74, 75, 85, 106
Reusch, William J. 242
▬▬▬▬▬ viii
Ritland, Osmund J., Maj. Gen. (USAF) 14, 19, 227
Root, Betty viii
Rosenau, Milton D. 44
Ross, B. Alan 243
Royce, Winston W. 242

SSS

S-2 Program 39, 42
Saint Program 13, 16
Samos Program 13, 15, 16, 19, 25, 229
Satellite basic assembly (SBA) 65, 67, 121, 237
Satellite control section (SCS) 81
Satellite control facility (SCF) 90, 129, 237
Satellite Operations Center (SOC) 90, 128
Satellite Test Center (STC) 90, 129
Schadegg, John J., Lt. Col. (USAF, retired) viii
Schlesinger, James R. 106
Schriever, Bernard A., Gen. (USAF) 7, 10, 14, 17
Scoville, Herbert, Jr. 20, 25
Secretary of the Air Force, Missiles & Space (SAFMS) 15, 19
Secretary of the Air Force, Special Projects (SAFSP) 15, 16, 17, 19, 25, 49, 50, 62, 109, 114, 235, 241
Sensor subsystem 65, 83
Sentry Program 13
▬▬▬▬▬ 109
Shea, Joseph 41
Sheldon, Huntington 40, 55, 64
Sheppard, William A., Col. (USAF)
Skantze, Lawrence 41
Smart, Jacob, Maj. Gen. (USAF) 48
Smith, Francis 242
Solar Array Module (SAM) 82
Solid-state-sensor (S3) 111, 114
Sorenson, Robert 44
Sorrels, John D. 243
▬▬▬▬▬ Field Operations Office 129
Space Systems Division (SSD) 17
Space Technology Laboratories (STL) 29, 41
Space Transportation System (STS) 237
Special Assistant to the President for Science and Technology 4, 232
Special Projects Staff (SPS) 38
Sputnick I and II 2, 8, 10
Steininger, Donald 61
Stellar index (SI) camera 64, 65, 75, 86, 106, 133
Stellar terrain camera 74, 86, 106, 133
Stewart, James T., Brig. Gen. (USAF) 35, 56
Strategic Air Command (SAC) 1, 15, 140
Strategic Arms Limitation Talks (SALT) 93, 136, 174
Strand, William, Col. (USAF) 35
Swenson, Lloyd S., Jr. 255
System Analysis Staff (CIA) 38
System Engineering and Technical Direction (SE/TD) 37
System operational requirement (SOR) 63, 65

TTT

TALENT-KEYHOLE Security System (TK) 124, 130
Tamanini, Elmer B. 242
Technological Capabilities Panel (TCP) 3, 4
Thompson-Ramo-Wooldridge (TRW) 25, 29, 32, 37, 73, 121, 242
Thor 2, 7, 9, 13, 227

Titan, Titan II, Titan III, Titan IIID 9, 32, 35, 38, 68, 76, 104, 121, 229
Tracking, telemetry, and command module (TTCM) 82
Transit (navigation satellite program) 12, 13
Treat, Steve P. 242
T'Unity 83, 90, 129

UUU

U-2 aircraft 3, 14, 15, 22, 23, 229
US Air Force 8, 11, 127
US Army 7, 11, 16, 127
US Navy 8, 11, 16, 127
Ultra-thin-base film 111
Ultra-ultra-thin-base film 111
United States Intelligence Board (USIB) 26, 127, 231

VVV

V-2 Rocket 7,8
Valley Program 39
Van Braun, Wernher 7
Vance, Cyrus 32, 34, 40, 54, 64, 235
Vandenberg, Hoyt S., Gen. (USAF) 8
Vandenberg Air Force Base 67, 95, 227
Vanguard 2, 8, 12
Velr-Hotel 16
Viking 8

WWW

Watson, John T. 242
███████ viii
Weber, Max 22, 24, 26
Weiss, Stanley I. vii, 68, 242
Welzenbach, Donald vii, viii
Werner, Richard W. 68
Western Development Division (WDD) 7, 9, 10, 13, 17
Wheelon, Albert D. 25, 27, 29, 31, 32, 39, 47, 55
White Sands Proving Ground 7, 8
Wiesner Report 16
Wilson, Charles E. 7
Wolfe, John 41
World aeronautical grid (WAG) 140, 145
Worthman, Paul E. viii, 43
Wright Air Development Command (WADC) 9, 17
WS-117L 9, 13, 227, 229

YYY

Yaksick, David M. 242

ZZZ

Zuckert, Eugene 17

Notes and References

1. Dwight D. Eisenhower, Mandate for Change 1953-1956, Doubleday and Co., Inc., Garden City, New York, 1963, p. 1.
2. Dwight D. Eisenhower, The White House Years: Waging Peace 1956-61, Doubleday and Co., New York, 1965, p. 390.
3. Welzenbach, Donald E., "Science and Technology: Origins of a Directorate," Studies in Intelligence, Vol. 30, No. 2, Summer 1986.
4. James R. Killian, Jr., Sputnik, Scientists and Eisenhower: A Memoir of the First Special Assistant to the President for Science and Technology, M.I.T University Press, Cambridge, Mass., 1977, p. 71.
5. Ibid., p. 82.
6. This assignment of a major "hardware" development program to a CIA executive was unique and is of particular interest in light of subsequent developments in the satellite reconnaissance program.
7. Augenstein, Bruno W., "Evolution of the U.S. Military Space Program, 1945-60," Fletcher School of Law and Diplomacy, Tufts University, 1982.
8. During this same period, Eisenhower stated that he had "come to regret deeply" that the missile program had not been set up in the Office of the Secretary of Defense, rather than in any of the military services. He said he would object emphatically to putting "the satellite job" in a military service. (Legislative Leadership Meeting, Supplementary Notes, 4 Feb 58, Dwight D. Eisenhower Diary Series, Staff Notes, February 1958, p.1, Eisenhower Library.)
9. Killian, op. cit., p. 127.
10. Lloyd S. Swenson, Jr., James M. Grimwood, Charles C. Alexander, This New Ocean, A History of Project Mercury, National Aeronautics and Space Administration, Washington, D.C., 1966, p. 29.
11. Killian, op. cit., p. 127.
12. Robert Perry, Origins of the USAF Space Program, AFSC Historical Publications Series 62- 24-10, p. 31.
13. Killian, op. cit., p. 128.
14. Swenson, Grimwood and Alexander, op. cit., pp. 76-7.
15. Named for a small Grecian Island. Almost instantly (and erroneously) elevated to acronym status by the press as representing "satellite and missile observation system."
16. Killian, op. cit., p. 82.
17. Itek is a company formed by former employees of the Boston University Optical Research Laboratory.
18. Work Statement, "Development of a Satellite Reconnaissance and Recovery System (CORONA) by LMSC," 25 Apr 58, COR-0016.
19. Later, when Ritland succeeded Schriever as Commander, AFBMD, he was succeeded in his CORONA role by Col. William A. Sheppard (also at AFBMD).
20. George B. Kistiakowsky, A Scientist at the White House, Harvard University Press, Cambridge, Mass., 1976, pp. 115 and 251.
21. For further information on CORONA, see Appendix B.
22. Kistiakowsky, op. cit., p. 382.
23. Taped interview, Air Force Secretary Eugene Zuckert, 25 Jul 64, by Lawrence McQuade, Kennedy Library, Boston, Mass.
24. Kistiakowsky, op. cit., p. 388.
25. The nation's intelligence operation coordinating board at that time.
26. Kistiakowsky, op. cit., pp. 47, 106, 202.
27. In the years that followed, the NRO's denigrators—particularly CIA staffers—chose to refer to the NRO as "the Air Force." This was a very clever usage: it made the NRO a victim of its own security. Even the (Kleyla) CIA history indulges the error, saying "[the agreement] gave the Air Force virtual control over all CIA programs." In later years, when the CIA addressed jurisdictional complaints to Secretary McNamara, it required less valor to impugn "the Air Force" than to say "your NRO."
28. H. H. Garth and C. Wright Mills, From Max Weber: Essays in Sociology, Oxford University Press, New York, 1946.
29. Ibid., p. 79.
30. Garth and Mills, op. cit., p. 54.
31. Ibid., p. 55.
32. Ibid., p. 196.
33. As a staff wag put it later, "If it hadn't been for Castro and Comsat, we would have had a beautiful NRO."
34. "Temporary" CORONA flew (very successfully) until May 1972.
35. Interview, Maj. Gen. John L. Martin, Jr., 26 Jan 88.
36. The United States Intelligence Board was an organization for formulating national intelligence requirements. See Annex C.
37. CIA Report, "Office of Special Projects 1965-1970" by ▓▓▓▓▓▓▓ approved by DDS&T June 1973, Vol. I, pp. 105, BYE-6400-72 TS.
38. Welzenbach, op. cit., p. 26.
39. DCI, Memo to DNRO, "CORONA Improvement Program," 22 Oct 63.
40. A. D. Wheelon, DDS&T, Memo for DDCI thru Executive Director-Comptroller CIA, Subject: Project FULCRUM, 23 Jun 64, BYE-0152-64.
41. Ibid.
42. Ibid.
43. NRO "Chronology of NRO-CIA Relationships," Vol. I, p. 51.
44. DDS&T Memo for DDCI, 23 Jun 64, op. cit.
45. Ibid.
46. DDR&E abbreviates Director of Defense Research and Engineering; Dr. Eugene Fubini, the Deputy Director, was designated to deal with NRO matters.
47. NRO Chronology, op. cit., Vol. I, p. 76.
48. CIA "Memorandum of Conversation," A. D. Wheelon, DDS&T, dated 18 Jun 64, subject, "Project FULCRUM—Satellite Photography System."
49. NRO Chronology, op. cit., pp. 76-78.
50. Ibid.
51. Ibid.
52. Memo for Record, A. D. Wheelon, DDS&T, 26 Jun 1964, subject: Land Panel Review of Project FULCRUM, BYE-0162-64.
53. Ibid.
54. It is not clear from the context whether Fubini meant that these were unknown causes of known defects or unknown defects.
55. For further information on GAMBIT, see Appendix C.
56. NRO Chronology, pp. 81-82.
57. Ibid., pp. 84-85.

SECRET
NOFORN-ORCON

58. In choosing STL as integrating contractor, Wheelon, a former STL employee, was recalling the company's strong capability gained as the system engineering contractor for the Air Force ballistic missile program.
59. NRO Chronology, op. cit., p. 87.
60. Ibid., p. 88.
61. DDS&T Memo to DDCI, 23 Jun 64, op. cit.
62. This purposefulness was also demonstrated in June 1964 by DDS&T's initiation of ▓▓▓▓▓▓▓▓▓▓▓▓▓▓▓▓▓▓▓▓▓▓▓▓▓▓▓▓▓▓ The courage of Wheelon's staff is noteworthy, inasmuch as only 11 months earlier, on 27 July 1963, Syncom-II, the first but very basic synchronous communication satellite, had been launched by NASA for the Comsat Corporation.
63. GAMBIT resolution in 1964 was generally between two and three feet. A subsequent major program change—GAMBIT-3—produced considerably better resolution than that envisaged for the follow-on search and surveillance system (FULCRUM).
64. NRO Chronology, op. cit., pp. 92-3.
65. Ibid., pp. 91-92.
66. This is with respect to FULCRUM. Other programs were discussed at the 11 August 1964 meeting. NRO Chronology, op. cit., pp. 94.
67. Ibid., p. 95.
68. Memo for Record, A. D. Wheelon, DDS&T, 18 Aug 64, subject: Meeting on 13 August 1964 with Dr. McMillan regarding Phase I Efforts on FULCRUM. All quotations in this paragraph are from this memo.
69. Chronology, op. cit., p. 98.
70. DDCI Memorandum to DDS&T, 27 Aug 64.
71. Memorandum for DCI thru DDCI, 31 Aug 64, Subject: Conduct of the FULCRUM Program.
72. NRO Chronology, op. cit., p. 108.
73. NRO Chronology, op. cit., pp. 127-128.
74. ExCom: Abbreviation for the NRO Executive Committee, consisting originally of the Deputy Secretary of Defense and the DCI with the later addition of the President's Science Advisor.
75. NRO Chronology, op. cit., p. 109.
76. A. D. Wheelon, DDS&T, Memo for DCI thru DDCI, 31 Aug 64, Subject: Conduct of the FULCRUM Program, BYE-0169-64 TS.
77. "Office of Special Projects 1965-1970" op. cit., Vol. 1, p. 114.
78. Crowley had been a senior addition to Wheelon's organization. One of his first assignments was a resolution of problems between SAFSP and CIA, at the working level, on CORONA.
79. Memo for Record, J. J. Crowley, SPS/DDS&T, 4 Dec 64, subject: "Visit to Martin/Denver with D. Rauth."
80. VALLEY was the name of SAFSP's general program for improving satellite photographic technology.
81. QPR, Program A, 30 Sep 65, BYE-40291-65.
82. Ibid.
83. J. J. Crowley, DSP, Memo to DDS&T, 29 Aug 66, Subject: Chronology of FULCRUM Effort with Itek, BYE-1649-66 S/C/H.
84. Incorrect in that SAFSP projects to study the optimum search and broad area coverage system payload paralleled or preceded CIA efforts.

85. A. D. Wheelon, DDS&T, Memo for DDCI, 8 Sep 64, Subject: NRO Reactions Stimulated by FULCRUM, BYE-0249-64,5.
86. Memo for the Record, A. D. Wheelon, DDS&T, 26 Jun 64, Subject: Land Panel Review of Project FULCRUM, op. cit.
87. NRO Chronology, op. cit., p. 100; A. D. Wheelon, DDS&T Memo for DCI, 30 Sep 64, Subject: Project FULCRUM, BYE-0282-64 S.
88. A. D. Wheelon, DDS&T, Letter to DNRO, 2 Oct 64, Subject: FULCRUM Program, BYE-0284-64, TS.
89. CIA Report, "Office of Special Projects 1965-1970," op. cit., p. 169.
90. NRO Chronology, p. 141.
91. NRO Chronology, op. cit., p. 143.
92. Loc. cit.
93. NRO Chronology, op. cit., p. 149.
94. Ibid., p. 151.
95. CIA Report, "Office of Special Projects 1965-1970," op. cit.
96. NRO Chronology, op. cit., p. 166.
97. CIA Report, "Office of Special Projects 1965-1970," op. cit.
98. Memorandum for Record, J. J. Crowley, SPS/DDS&T, 19 Jan 65, subject: FULCRUM Status Briefing BYE-0027-65.
99. NRO Chronology, op. cit., p. 166.
100. Ibid., pp. 160-161.
101. A subtle but significant expansion of STL's role, which may have been a contributor to strained CIA-Itek relationships.
102. NRO Chronology, op. cit., pp. 163-164.
103. Ibid., p. 167.
104. The "1/4" was the ratio of the folded part of the optics (per W. Levison in a 22 Jan 88 interview).
105. "Chronology of FULCRUM Effort With Itek," op. cit., par. 12.
106. Ibid., par. 15.
107. Ibid., par. 16.
108. Ibid., par. 18.
109. Interview, Walter Levison, 22 Jan 88.
110. Interview, Levison, op. cit.; NRO Chronology, op. cit. p. 173.
111. Interview, Levison, op. cit.
112. Ibid.
113. Ibid.
114. NRO Chronology, op. cit., p. 170.
115. Ibid., p. 171.
116. Interview, Levison, op. cit.
117. NRO Chronology, op. cit., p. 172.
118. Ibid., p. 173.
119. Ibid., p. 174.
120. "Office of Special Projects, 1965-1970," op. cit., p. 173; "Chronology of FULCRUM Effort With Itek," op. cit., par. 24.
121. QPR, Program A, 1 April-30 June 1965.
122. QPR, Program A, 1 July-30 September 1965.
123. QPR, Program A.
124. Perkin-Elmer Co., "A History of the HEXAGON Program 1985," BIF007-0235-85.
125. Interview, John McMahon, August 1987.
126. Perkin-Elmer History, op. cit., p. 7.
127. Ibid., p. 8.

SECRET
Handle via
BYEMAN-TALENT-KEYHOLE
Control Systems Jointly
BYE 140003-92

128. NRO Chronology, op. cit., p. 192.
129. Ibid., pp. 193-4.
130. Ibid., pp. 194-5.
131. Ibid., p. 195.
132. Memo for DDCI from DDS&T, 25 Feb 65, Subject: "Establishment of a Satellite Office Within the Science and Technology Directorate," Secret, BYE-0089-65.
133. Memo, R. M. Bissell to Maj. Gen. J. R. Smart, Dep. V/C Staff, USAF, Subject: "Distribution of Responsibilities for CORONA," 25 Nov 58, SAFSS files.
134. Kistiakowsky, op. cit., p. 106.
135. Memo, J. A. Cunningham, Dep. Asst. Dir. (Special Activities) to Exec. Dir. CIA, Subject: "Basic Authorities and Agreements which CIA holds with USAF/NRO," 24 Jan 63.
136. "A Summary of the National Reconnaissance Program," A. D. Wheelon, 13 May 65, BYE- 0204-65, Top Secret.
137. "Chronology of NRO-CIA Relationships," Vol. II, p. 206.
138. Loc. cit.
139. The implication is that the Executive Committee (ExCom) did not exist before this time. In actuality the group had functioned on an informal basis, without charter, for the previous year. It was a natural outgrowth of the NRP Review Committee established in December 1963.
140. NRO Chronology, op. cit., p. 208.
141. See Appendix D.
142. Ibid.
143. The Agreement does not mention SIGINT programs such as ▓▓▓▓ which was formally approved about this time and remained assigned to the CIA. It is likely that its concentration on optical systems reflected the problem existing between DoD and CIA at the time.
144. Appendix D, op. cit.
145. Ibid.
146. In early 1965, the ad hoc Land Panel was reconstituted as a permanent body under the PSAC.
147. Memo for Deputy Secretary of Defense from D. F. Hornig, Special Assistant to the President for Science and Technology, 30 Jul 65, BYE-58332/65.
148. NRO Chronology, op. cit., p. 219.
149. Ibid., p. 219.
150. "Office of Special Projects, 1965-1970," op. cit. Vol. I, p. 125.
151. Communication ▓▓▓▓ November 1988.
152. Interview, Dr. A. Flax, August 1987.
153. NRO Chronology, op. cit., p. 222.
154. Ibid., p. 224.
155. Ibid.
156. Ibid.
157. Ibid.
158. FULCRUM Development Summary and Progress, 1 October 1965 to 31 March 1966, (S), 22 Apr 66; Memorandum for Program Director, HEXAGON from J. J. Crowley Dir/SP, 16 Feb 68, Subject: Updating Program History, HEX-4701-68.
159. NRO Chronology, op. cit. pp. 229-31.
160. Draft, A. D. Wheelon, 2 Nov 65, BYE-0455-65.
161. NRO Chronology, op. cit., pp. 231-233.
162. Ibid., p. 233.
163. DSPD was the designation used for a Deputy System Program Director. In this case, it referred to the person in charge of the payload, a management function that both Generals Martin and Stewart believed should be located with the rest of the system management team in Los Angeles.
164. NRO Chronology, op. cit., p. 234.
165. Office of Special Projects, 1965-1970, op. cit., p. 178; Interview, Dr. A. Flax, August 1987, op. cit.
166. Flax interview, op. cit.
167. It subsequently developed that the code name HELIX had previously been assigned to another project; accordingly the search program was soon redesignated HEXAGON.
168. Memorandum from DNRO to Maj. Gen. Martin, Sheldon, Steininger, 1 Apr 66, Subject: "New General Search and Surveillance System."
169. Communication from Brig. Gen. David L. Carter, USAF (Ret) Mar 88.
170. The three were the Perkin-Elmer design, which came from FULCRUM, and the two ITEK designs, which came from S-2 (one of them originally pursued by Eastman Kodak).
171. NRO Chronology, op. cit. p 244a, b.
172. TWX to Flax from Martin, ▓▓▓▓ 5 Apr 66.
173. NRO Chronology, op. cit., pp. 244b, c.
174. H. D. Sheldon, Dir/Recon, CIA, Memo to DNRO, 7 Apr 66, Subject: HELIX Program, BYE-0075- 66, TS.
175. NRO Chronology, op. cit., pp. 246-247.
176. Between the time the DNRO submitted his formal input to the ExCom on 22 April and the ExCom meeting on 26 April, the name was changed from HELIX to HEXAGON.
177. NRO Chronology, op. cit., p. 248.
178. Ibid., p. 249.
179. Ibid.
180. Ibid.
181. Attachment to ltr, DNRO to Dir SAFSP and Dir Reconnaissance CIA, 29 Apr 66.
182. Ibid.
183. The terms "recovery vehicle" and "reentry vehicle" refer to the same system component and are used synonymously.
184. The Stellar Index Camera eventually, with additional capabilities, became the Stellar- Terrain Camera.
185. Attachment 3, DNRO Memo to ExCom regarding New General Search and Surveillance Satellite System, March 1966, BYE-52224-66.
186. The Sensor Subsystem Request for Proposal did not specify dimensions. The original FULCRUM diameter was 90 inches, as proposed for HEXAGON by Perkin-Elmer. This dimension was carried by the SSPO as a requirement in OSP (SETS) document IRD-501 "AVE Interface Requirements for Sensor Subsystem" 31 Mar 67, updated to 13 Nov 67, despite the fact that the SBA had decided on a diameter of 120 inches. After the Camera Frame Preliminary Design Review, 23 May 1967, the SSPO sought and received relief to 100 inches for the sensor package.
187. IRD-501, op. cit.
188. Perkin-Elmer HEXAGON Program Proposal, 21 Jul 66, Vol II, pp. 2-9, BYE-1671-66 (TS).
189. Ibid.

SECRET
NOFORN-ORCON

190. ███████, "HEXAGON History," (draft), 29 Sep 73, BYE-107859-73.
191. ███████ History, op. cit.
192. Ibid.
193. Memo, Subject: "Management Plan for the HEXAGON Project," ███████ to D/OSP, 3 May 66, BYE-1572-66.
194. ███████ History, op. cit.
195. QPR, Program A, 1 April-30 June 1968.
196. Memo, "NPIC's Attitude Accuracy Requirements for HEXAGON", from A. C. Lundahl, D/NPIC, 15 Aug 68, BYE-2966-68 (TS).
197. Report, HEXAGON (KH-9) Mapping Camera Program and Evolution, December 1982, prepared for SAFSP, BIF-059W-23422/82, pp. 2-5 to 2-7.
198. Request for Proposal, Satellite Basic Assembly (SBA) for the Photographic General Search and Surveillance Satellite System (HEXAGON), distributed 16 Jun 66.
199. "LMSC Acquisition Phase Proposal, Program HEXAGON," Vol II, 22 Aug 66.
200. IRD No. 501, op. cit.
201. ███████ History, op. cit.
202. A proven concept derived from earlier satellite systems design of CORONA and GAMBIT.
203. The material from this point to the section "The Donovan Review Committee" is taken from "A History of the HEXAGON Program" BIF007-0253-85 by Richard J. Chester (Perkin-Elmer Corporation).
204. Later availability of ultra-ultra-thin-base (UUTB) film increased this capacity to 155,000 feet.
205. HEXAGON (KH-9) Mapping Camera Program and Evolution, BIF-059W-23422/82, Maurice Burnett, editor.
206. Ltr, Martin to A. Donovan, Aerospace Corporation, 18 Oct 68.
207. QPR, 31 Mar 69.
208. Perkin-Elmer History, p. 134.
209. Robert Perry, "A History of Satellite Reconnaissance," Vol IIIB, BYE-17017-74, pp. 81-82.
210. ███████ History.
211. Communication, Don Welzenbach, November 1988.
212. Ibid.
213. Ibid., p. 86.
214. A History of the HEXAGON Program (Perkin-Elmer), BIF007-0253-85.
215. Report, "System Performance Evaluation Team, Mission 1201," 27 Sep 71, BYE-15285-71 (TS).
216. Report, "Performance Evaluation Team, Mission 1201," 27 Sep 71, BYE-15285-71 (TS).
217. Perkin-Elmer History, p. 182.
218. Interview, Henry Epple, 8 Apr 88.
219. QPR, Program B, 1 Jan-31 Mar 72, BYE-6410-72.
220. Lifeboat-II was a back-up and separate subsystem, independently ground-commandable, to orient the SV so that an RV could be separated and the SV then propelled to reenter and impact in a designated remote ocean area.
221. As a part of each mission, a "solo" phase was normally conducted (after the RVs had been separated and reentered) during which various engineering tests were conducted while the hardware was still in space.

222. Report, "System Performance Evaluation Team, Mission 1202," 21 Apr 72, BYE-15265-72 (TS).
223. Report, "System Performance Evaluation Team, Mission 1203," 6 Nov 72, BYE-15310-72 (TS).
224. Ibid.
225. Ibid.
226. Ibid.
227. Report "System Performance Evaluation Team, Mission 1204," 23 Feb 73, BYE-15260-73. Note that a later Perkin-Elmer summary report PM-1521-X-H, 11 May 81, BIF007-0881-74H (revolution) quotes a ███████ average of 3.4 feet.
228. Ibid.
229. Ibid.
230. The Performance Evaluation Team that prepared the cited reports on system performance.
231. Report, "System Performance Evaluation Team, Mission 1205," 4 Sep 73, BYE-15296-73.
232. Report, HEXAGON (KH-9) Mapping Camera Program and Evolution, December 1982, BIF-059W-23422/82.
233. Report, "System Performance Evaluation Team, Mission 1206," 5 Feb 74, BYE-15323-73 (TS).
234. TWX ███████ 291940Z Nov 71; TWX ███████ 021057Z; TWX ███████ 032140Z Dec 71.
235. Ibid.
236. HEXAGON Transition Plan, March 1972, BYE-93603-72.
237. It assumed that SV-6 would be flown by the transition date, and only SV-6 postflight analysis, fee determination, and contract closeout actions would remain open.
238. Transition Plan, op. cit.
239. Memo for D/OSP, ███████ OSP, 24 Feb 72, "HEXAGON Program Transfer to the Air Force."
240. TWX ███████ 3020212, June 1973.
241. TWX ███████, 072255Z August 1976.
242. Report, "System Performance Evaluation Team, Mission 1207," August 1974, BYE-15319-74.
243. "HEXAGON Program Preliminary Postflight Report For Flight No. 7," 5 Mar 74, BIF-107W- 71001-74.
244. ███████
245. This Block III change was flown early beginning with Vehicle 11. See "Performance Evaluation Team Report, Mission 1211," November 1976, BYE-3225/76.
246. "Performance Evaluation Team Report, Mission 1213," June 1979, BYE-9570/79.
247. These changes were effective on SV-17; Report: "A History of the HEXAGON Program," 1985, BIF007-0253-85.
248. "Performance Evaluation Team Report, Mission 1214," February 1980.
249. Letter from Col. F. J. Haber, Dep Cmdr/Satellite Operations to Cmdr, AFSCF, 24 Oct 79, Subject: "Operations Evaluation Report for IRON 3854" (S).
250. Sensor System Postflight Report, Mission 1215 (S/N 018) PM-1689-X, 8 Jan 80, BIF007/D- 0203-80.
251. Postflight Analyses Report, Mission 1216-5, 27 Mar 81, BIF-059W-23016-81.
252. HEXAGON Program Preliminary Postflight Report for Flight No. 17, 14 Jan 83, BIF-107W- 12002-83 (S/H) p. 9.
253. Loc. cit.

SECRET
Handle via
BYEMAN-TALENT-KEYHOLE
Control Systems Jointly
BYE 140003-92

SECRET
NOFORN-ORCON

254. Ibid., pp. 10, 78.
255. HEXAGON Program Preliminary Postflight Report for Flight No. 18, 14 May 84, BIF-0107- 12009-84 (S/H).
256. Operations in which the solid-state-sensor (S3) subsystem was used for precise attitude determination.
257. SCAD — S-Cubed Activity Detection.
258. HEXAGON Program Preliminary Postflight Report for Flight No. 19, 14 Dec 84, BIF-0107- 12015-84 (S/H) p. 6.
259. Ibid., p. 6.
260. Ibid., p. 7.
261. HX Financial History, ████, 28 Jul 88, BYE-111225-85 (TS).
262. HEXAGON flew a total of 2,258 photographic days.
263. Based upon available data, HEXAGON covered ████ unique ████ targets.
264. Similarly, HEXAGON covered approximately 150 million unique cloud-free square nautical miles.
265. NSC "US Policy on Outer Space Including Satellite Activity," 10 Jul 62 (TS).
266. Eisenhower, "Waging Peace, 1956-1961," Doubleday and Company, Inc., Garden City, N.Y., 1965, p. 467.
267. Director of Central Intelligence Directive No. 1/13, Effective 1 Jul 67.
268. USIB-D-41-14/224 (COMOR-0-13/63) 21 Jun 66.
269. TK System Identifiers used in this document were KH-4 (CORONA), KH-7 (GAMBIT) and KH-9 (HEXAGON).

270. As the number of missions flown was reduced and their on-orbit duration greatly extended, this requirement was dropped. The effectiveness of long-duration missions, in terms of requirements satisfaction, outweighed the benefits of more current imagery. Other programs satisfied most of the current imagery requirements, especially after 1976.
271. The R-3 standby capability was later dropped when the NRO reported it would cost on the order of ████ dollars to construct a new launching pad. The NRO noted it could maintain a backup vehicle capability at about R-9 days with the single launching pad.
272. USIB-D-41-14/296, 20 Jul 66.
273. USIB-D-41-14/295, 11 Jul 66.
274. USIB-D-41-14/294 (COMOR-D-13/63) 21 Jun 66.
275. USIB-D-46-4/32, 10 Nov 69.
276. The imagery Satellite Search Performance Study (SPS) BYE-2249, July 1975.
277. Assessment of the Intelligence Gain Provided by KH-9 over KH-4 and KH-8, COMIREX D-11- 1/2, 1 Jul 68.
278. Final Report of Design Study for HEXAGON Reconnaissance System using STS (TS), LMSC, 31 Aug 73, BIF003W/2-069331-73.
279. Ibid.
280. Ibid.
281. Last HEXAGON flight (vehicle 20).
282. Responsibility transferred to SAFSP effective this date.

SECRET
Handle via
BYEMAN-TALENT-KEYHOLE
Control Systems Jointly
BYE 140003-92

www.ingramcontent.com/pod-product-compliance
Lightning Source LLC
Chambersburg PA
CBHW082036300426
44117CB00015B/2509